The Quotable Soldier

The Quotable
Soldier

COMPILED BY
LAMAR UNDERWOOD

THE LYONS PRESS

FIRST EDITION

Printed in the United States of America

10 9 8 7 6 5 4 3 2 1

Library of Congress Cataloging-in-Publication Data is available on file.

DEDICATED

WITH DEEP GRATITUDE AND RESPECT

TO ALL THE MEN AND WOMEN,

LIVING AND DEPARTED,

WHO HAVE SERVED

IN THE ARMED FORCES

OF THE UNITED STATES OF AMERICA.

AND TO THE MEMORY OF MY FATHER,

LT. COL. JOHN DOUGLAS UNDERWOOD

(U.S. ARMY, RETIRED)

It's payback time, Dad, from the luckiest "Army Brat" of them all

Order of Battle

In the final choice, a soldier's pack
is not so heavy a burden as a prisoner's chains.

PRESIDENT DWIGHT D. EISENHOWER
FIRST INAUGURAL ADDRESS, 1953

Introduction

Every man thinks meanly of himself for not having
been a soldier.

SAMUEL JOHNSON, 1778

He knew a thing or two, that Samuel Johnson. He could
fling barbs as well as any man, and this particular one was
hurled dead-on at his biographer, James Boswell. According
to the noted military historian John Keegan, Johnson knew
that Boswell as a young man in Britain had secretly longed
to serve in the Foot Guards and wear an officer's scarlet coat.
The wise-cracking Johnson himself had no such ambitions,
but he understood Boswell's harmless fantasy, and he knew
that many men who had never felt the sting of battle or
known a single moment of the drudgery of service life har-
bored deep regrets: No matter how full or successful their
lives had been, they missed never having been part of some-

thing important and unforgettable—part of a long line of men who had marched through the pages of history and, one way or the other, did whatever they could to serve their comrades and their country.

Not everyone, of course, is so sentimental about military service. On the contrary, there are those who loathe the subject, even though they have been in uniform. Some loathe it *because* they have been in uniform. And then there are the sanctimonious, professional "peacemakers" babbling to everyone what a terrible mistake war is (as if they have discovered a unique and illuminating vision), meanwhile bunkering themselves deep into comfortable and safe positions while letting others tend to the fighting.

I suspect things have always been that way, although in today's atmosphere of instant information peppered with hype and salted with spin-doctoring, integrity and commitments to *anything* sometimes seem as fragile as peace itself.

Some veterans of bitter combat never speak of their experiences—ever!—and their Purple Hearts and other medals are tucked deep in their dresser drawers. Others proudly doff their VFW caps to march in parades, and still others pin medals and service patches on old jungle camo shirts when they visit that long dark wall beside the Potomac River where they can reach out and touch the name of someone very special who never made it home.

The horrors of war and combat are so obvious that our fascination with the subject must be questioned: Why? Why the books, the monuments, the parades, the films, the TV documentaries, the speeches, the endless running arguments in the op-ed pages? Wouldn't you think that the anguish and sheer heartbreak of the wars of the twentieth century alone would have resulted in a condition of silent numbness?

The answer lies, I believe, in the fact that we do not *want* to forget. As the distinguished novelist Herman Wouk reminds us, "The beginning of the end of War lies in Remembrance."

Whether we are veterans with distinguished records of service, or amateur military history buffs who have never even pulled KP or slogged through boot camp (like myself), it seems important to remember the voices and deeds of soldiers, marines, sailors, and airmen of the past. When we listen carefully, they speak to us across the gulf of centuries and decades of what they saw, what they felt.

Call me old-fashioned, a sentimental fool, if you wish, but include me among those who want to know what it was like at Valley Forge, Gettysburg, Belleau Woods, Tarawa and Iwo Jima, Normandy Beach, the Chosin Reservoir, Khe Sanh, and Kuwait. I want to know what the jokes were, and the songs the troops sang, and what they said in the letters and books they went on to write. I want to know what the politicians were telling them, and what the generals were telling them, and what they were telling each other.

In the pages ahead, the obvious limitations of space prevent us from providing the details that would make this a definitive book on soldiering through war and peace. The books that achieve that objective are many, and we have quoted and recommended several here, including some of the very newest—like Martin Russ's *Breakout*, the story of the Marines fighting their way out of the Chosin Reservoir in Korea; and Tom Clancy's *Every Man a Tiger*, the story of General Chuck Horner and air power from Vietnam through the Gulf War. Our quotations, however, are worthy little nuggets of pure military history gold. Within them lie the heartbeats and, if you will excuse me, the blood, sweat, and tears that lurk beneath the facts and dates that reside in history books.

Veterans and military history buffs do not spend all their time reminiscing tearfully about courageous men and women and their actions, or arguing over such weighty issues as why Montgomery was so slow to take Caen, or what would have happened to Wellington at Waterloo had not Blucher's reinforcements arrived. Remembered also are the lighter moments that break the frequent stretches of utter boredom in military life. Some of the more memorable of these are here, and hopefully we have not left out one of your favorites.

Another of my aims in assembling this book was to provide what I hope will be a handy and accurate little source of settling trivia arguments, the kind that military history buffs are wont to engage in: How many times did Longstreet recommend to Lee that their army swing to the right to encircle Meade's Union forces? Who was Colonel Paul Tibbets? What was his mother's name? Military folk love this stuff, and I trust I've got it right for you—especially when you've got a bet down!

In gathering these quotes into some kind of interesting format that makes sense, I immediately thought that a straight chronological format would be dull, and I also rejected an impulse to list them by battle for the same reason. Instead, I am presenting them in the way I feel will be the most readable, in various categories representing different phases of military life. In two cases, Pearl Harbor and Vietnam, I have chosen to gather the quotes into distinctive, individual sections—for reasons I trust will become obvious as you read.

Some of the quotes have certain attributes that could have qualified them for being placed in any one of several categories. The call was mine to make, and that's what I did.

Thanks for listening.

Lamar Underwood
May 2000

The Quotable Soldier

1

The Call to Battle

Fighting Words—
of Inspiration, Anger, and Commitment

It is not set speeches at the moment of battle that render soldiers brave. The veteran scarcely listens to them, and the recruit forgets them at the first discharge. . . .

NAPOLEON I
Maxims, No. LXI (1831)

Napoleon was right, no doubt, but the words keep coming—always have, always will—and some of them aren't half bad or even boring. Lincoln's suggestion that "the world will little note nor long remember what we *say* here," certainly applies to most patriotic speeches and Knute Rockne–style cheerleading on the eve of battle.

But some of the voices men and women in uniform have heard and respected vibrate with a resonance and spirit that transcends time. Whether the entreaties of politicians, presidents, kings, or generals, or angry shouts heard over the static of a field radio or from some adjacent foxhole, certain words and phrases have burned a brand-mark in history and memory.

Most veterans will probably tell you that Ernest Hemingway got it right when he said, "Abstract words such as glory, honor, courage or hallow were obscene beside the concrete names of villages, the numbers of roads, the names of rivers, the numbers of regiments, and the dates."

Indeed, deeds—not words—win battles and wars. But sometimes, the words that preceded or followed the deeds are so strong that we just can't let go of them. In my opinion, those that follow certainly qualify for that distinction.

I hear a lot of crap about what a glorious thing it is to die for your country. It isn't glorious, it's stupid! You don't go into battle to die for *your* country, you go into battle to make the other bastard die for *his* country.

GENERAL GEORGE S. PATTON, JR.
Speaking to troops, 1941

Stand your ground. Don't fire unless fired upon, but if they mean to have a war let it begin here!

CAPT. JOHN PARKER
To his seventy-seven Minute Men on the village green at Lexington, Massachusetts, April 19, 1775.

Either come back from battle carrying your shield or being carried on your shield.

CREED OF THE SPARTAN SOLDIERS
(480 B.C.)

C'est la guerre!
(It is war!)

POPULAR FRENCH SAYING

I urge you to fly to arms, and smite with death the power that would bury the government and your liberty in the same hopeless grave.

FREDERICK DOUGLASS
"MEN OF COLOR, TO ARMS" (1863)

War is a matter of vital importance to the State; the province of life or death; the road to survival or ruin. It is mandatory that it be thoroughly studied.

SUN TZU
THE ART OF WAR (500 B.C.)

Then out spake brave Horatius,
The Captain of the Gate:
"To every man upon this earth
Death cometh soon or late.
And how can man die better
Than facing fearful odds,
For the ashes of his fathers,
And the temples of his Gods?"

THOMAS BABINGTON MACAULAY
LAYS OF ANCIENT ROME (1842)
One of Winston Churchill's favorite verses

I have nothing to offer but blood, toil, tears and
sweat.

WINSTON CHURCHILL
First statement as Prime Minister, House of Commons
(May 1940)

We shall go on to the end. We shall fight in France, we shall fight on the seas and oceans, we shall fight with growing confidence and growing strength in the air, we shall defend our island, whatever the cost may be, we shall fight on the beaches, we shall fight on the landing grounds, we shall fight in the fields and on the streets, we shall fight in the hills, we shall never surrender.

WINSTON CHURCHILL
Dunkirk Speech, June 4, 1940

He [Winston Churchill] mobilized the English language and sent it into battle to steady his fellow countrymen and hearten those Europeans upon whom the long dark night of tyranny had descended.

EDWARD R. MURROW (1954)

Once more into the breach, dear friends, once
    more
Or close the wall up with our English dead!
In peace there's nothing so becomes a man
As modest stillness and humility;
But when the blast of war blows in our ears,
Then imitate the action of the tiger;
Stiffen the sinews, summon up the blood,
Disguise fair nature with hard-favor'd rage;
Then lend the eye a terrible aspect.
. . . .

That he which hath no stomach to this fight,
Let him depart; his passport shall be made
And crowns for convoy put into his purse:
We would not die in that man's company
That fears his fellowship to die with us.
. . . .

This day is call'd the feast of Crispian:
He that outlives this day, and comes safe home,

Will stand a tip-toe when this day is named,
And rouse him at the name of Crispian.
He that shall live this day, and see old age,
Will yearly on the vigil feast his neighbours,
And say, "Tomorrow is Saint Crispian":
Then will strip his sleeve and show his scars,
And say, "These wounds I had on Crispin's day."
. . . .
We few, we happy few, we band of brothers;
For he today that sheds his blood with me
Shall be my brother; be he ne'er so vile
This day shall gentle his condition:
And gentlemen in England now a-bed
Shall think themselves accursed they were not here,
And hold their manhoods cheap whiles any speaks
That fought with us upon Saint Crispin's day.

KING HENRY
in WILLIAM SHAKESPEARE
KING HENRY THE FIFTH (1599)

When it's all over and you're home once more, you can thank God that twenty years from now, when you're sitting around the fireside with your grandson on your knee and he asks you what you did in the war, you won't have to shift him to the other knee, cough, and say, "I shoveled shit in Louisiana."

GENERAL GEORGE S. PATTON, JR.
To troops during D-Day buildup in England, 1944

War is not merely a political act, but also a political instrument, a continuation of political relations, a carrying out of the same by other means.

GENERAL CARL VON CLAUSEWITZ
ON WAR (1832)

Soldiers, Sailors and Airmen of the Allied Expeditionary Force:

You are about to embark on the Great Crusade, toward which we have striven these many months. The eyes of the world are upon you. The hopes and prayers of liberty-loving people everywhere march with you.

Your task will not be an easy one. Your enemy is well trained, well equipped and battle-hardened. He will fight savagely.

But this is the year 1944! . . . The tide has turned. The free men of the world are marching together to Victory!

I have full confidence in your courage, devotion to duty and skill in battle. We will accept nothing less than full victory!

GENERAL DWIGHT D. EISENHOWER
Order of the day handed to troops boarding transports and landing craft for D-Day, June 6, 1944

We were born naked and have been taught to hunt and live on the game. You tell us that we must learn to farm, live in one house, and take on your ways. Suppose the people living beyond the great sea should come and tell you that you must stop farming and kill your cattle and take your houses and lands, what would you do? Would you not fight them?

> INDIAN WARRIOR GALL
> Speaking to government commissioners as a young man. He was to become Sitting Bull's lieutenant.

Fight, gentlemen of England! Fight, bold yeomen!
Draw, archers, draw your arrows to the head!
Spur your proud horses hard, and ride in blood;
Amaze the welkin with your broken staves!

> KING RICHARD
> in WILLIAM SHAKESPEARE
> RICHARD THE THIRD (1593)

It is an unfortunate fact that we can secure peace only by preparing for war.

> JOHN F. KENNEDY
> Speaking as a presidential candidate, 1960

I offer neither pay, nor quarters, nor food; I offer only hunger, thirst, forced marches, battles and death. Let him who loves his country with his heart, and not merely with his lips, follow me.

> GIUSEPPE GARIBALDI (1849)
> Italian patriot and soldier

For courage mounteth with occasion.

> AUSTRIA
> in WILLIAM SHAKESPEARE
> *KING JOHN* (1596)

The dogmas of the past are inadequate to the stormy present. The occasion is piled high with difficulty, and we must rise with the occasion.

ABRAHAM LINCOLN
Second Annual Message to Congress, 1862

Forever, and forever, farewell, Brutus;
If we do meet again, why, we shall smile;
If not, why then, this parting was well made.

CASSIUS
in WILLIAM SHAKESPEARE
JULIUS CAESAR (1599)

Cry "Havoc!" and let slip the dogs of war.

MARK ANTHONY
IN WILLIAM SHAKESPEARE
JULIUS CAESAR (1599)
Addressing the people after Caesar's assassination.

———•••◦•••———

The Nazis entered this war under the rather childish delusion that they were going to bomb everybody else, and that nobody was going to bomb them. At Rotterdam, London, Warsaw and half a hundred other places they put that rather naïve theory into operation. They sowed the wind, and now they are going to reap the whirlwind.

AIR CHIEF MARSHAL ARTHUR HARRIS
Speaking as Commander in Chief, R.A.F. Bomber
Command (1943)

Tonight you go to the big city [Berlin]. You have the opportunity to light a fire in the belly of the enemy and burn his black heart out.

> AIR CHIEF MARSHAL ARTHUR HARRIS
> Message to crews as Commander in Chief, R.A.F.
> Bomber Command (1943)

My son, seek thee out a kingdom equal to thyself; Macedonia has not room for thee.

> PHILIP OF MACEDON
> Father of Alexander the Great

If we are marked to die, we are enough
To do our country loss, and if to live,
The fewer men, the greater share of honour.

> KING HENRY V
> in WILLIAM SHAKESPEARE
> *KING HENRY THE FIFTH* (1599)

This England never did, nor never shall,
Lie at the proud foot of a conqueror.

> PHILIP THE BASTARD
> in WILLIAM SHAKESPEARE
> *KING JOHN* (1596)

We go to gain a little patch of ground
That hath in it no profit but the name.

> HAMLET, PRINCE OF DENMARK
> in WILLIAM SHAKESPEARE
> *HAMLET* (1601)

Lafayette, nous sommes ici.
Lafayette, we are here.

> COLONEL CHARLES E. STANTON
> At the Tomb of Lafayette in Paris, July 4, 1917,
> referring to the arrival of the American Expeditionary
> Forces in France

These are the times that try men's souls. The summer soldier and the sunshine patriot will, in this crisis, shrink from the service of his country; but he that stands it *now*, deserves the love and thanks of man and woman. Tyranny, like hell, is not easily conquered; yet we have this consolation with us, that the harder the conflict, the more glorious the triumph.

THOMAS PAINE
THE AMERICAN CRISIS, No. 1 (1776)

They shall not pass.

MARSHAL HENRI PHILIPPE PETAIN
Referring to the attacking Germans during the French defense at Verdun (1916).

Ye shall hear of wars and rumors of wars.

> Matthew 24:6
> *BIBLE, NEW TESTAMENT*

———

Weary days with wars and rumors of war, and marching of troops, and flags waving, and people talking.

> HENRY WADSWORTH LONGFELLOW
> *JOURNAL* (1861)

———

Duty, Honor, Country

> U.S. MILITARY ACADEMY MOTTO

———

This aggression will not stand.

> PRESIDENT GEORGE H. W. BUSH
> After Saddam Hussein's invasion of Kuwait, 1990

The die is cast.

JULIUS CAESAR
At the crossing of the Rubicon

———•∙∙•———

Conscription may form a great and admirable machine, but it differs from the trained army of volunteers as a body differs from a soul. But it costs a country heavy in griefs, does a volunteer army; for the flower of the country goes.

MARY ROBERTS RINEHART
KINGS, QUEENS AND PAWNS (1915)

Day by day fix your eyes upon the greatness of Athens, until you become filled with the love of her; and when you are impressed by the spectacle of her glory, reflect that this empire has been acquired by men who knew their duty and had the courage to do it.

THUCYDIDES (471–401 B.C.)
"FUNERAL SPEECH OF PERICLES"
PELOPONNESIAN WAR, BOOK II

——◦•◦——

I am the Infantry—Queen of Battle! For two centuries I have kept our Nation safe, purchasing freedom with my blood. To tyrants, I am the day of reckoning, to the suppressed, the hope for the future. Where the fighting is thick, there am I. I am the Infantry! FOLLOW ME!

I was there from the beginning, meeting the enemy face to face, will to will. My bleeding feet stained the snow at Valley Forge; my frozen hands pulled Washington across the Delaware . . .

Where brave me fight, there fight I. In freedom's cause, I live, I die. From Concord Bridge to Heartbreak Ridge, from the Arctic to the Mekong, to the Carribbean, the Queen of Battle! Always ready, then, now and forever.

I am the Infantry! FOLLOW ME!

> CREED OF THE U.S. ARMY INFANTRY CENTER,
> FORT BENNING, GEORGIA

If the trumpet gives an uncertain sound, who shall prepare himself to the battle?

> I Corinithians 14:8
> *BIBLE, NEW TESTAMENT*

God is always on the side of the big battalions.

VOLTAIRE (1770)

We are on God's side.

JOE LOUIS (c. WWII)
The boxing champion answering a reporter's question,
"Is God on our side?"

There never was a good war or a bad peace.

BENJAMIN FRANKLIN (1775)

Remember the Maine!

ANONYMOUS
Slogan during the Spanish-American War (1898)

Over there, over there
Send the word over there
Yes, the yanks are coming
The drums rum-tumming everywhere
And we won't be back till it's over
over there.

> GEORGE M. COHAN (1917)
> "Over There"
> Popular WWI song

From the halls of Montezuma,
To the shore of Tripoli,
We fight our country's battles
On the land as on the sea.

> U.S. MARINES HYMN

Over hill, over dale, we have hit the dusty trail
And those caissons go rolling along.
Countermarch! Right about!
Hear those wagon soldiers shout
While those caissons go rolling along.
Oh, it's hi-hi-ye! For the field artilleree,
Shout out your numbers loud and strong,
And where're we go, you will always know
That those caissons are rolling along.

MAJOR EDMUND L. GRUBER
"THE CAISSON SONG" (1908)

Let Bacchus' sons be not dismayed
But join with me each jovial blade;
Come booze and sing and lend your aid
To help me with the chorus.
[Chorus]
Instead of Spa we'll drink down ale.
And pay the reck'ning on the nail;
No man for debt shall go to gaol
From Garry Owen in glory.

"GARRY OWEN" (C. 1800)
SONG OF THE 7TH CAVALRY

Anchors Aweigh, my boys, Anchors Aweigh.
Farewell to foreign shores, We sail at break of day-ay-ay.
Through our last night on shore, Drink to the foam,
Until we meet once more. Here's wishing you a
happy voyage home.

"THE NAVY MARCH" (1906)
LT. CHARLES A. ZIMMERMAN (MUSIC) AND
MIDSHIPMAN ALFRED H. MILES (LYRICS)

Eternal Father, Strong to save,
Whose arm hath bound the restless wave,
Who bid'st the mighty Ocean deep
Its own appointed limits keep;
O hear us when we cry to thee,
for those in peril on the sea.

"THE NAVY HYMN" (C. 1860)
REV. WILLIAM WHITING (1825–1878) AND REV. JOHN
DYKES (1823–1876)

Mine eyes have seen the glory of the coming of the
Lord:
He is tramping out the vintage where the grapes of
wrath are stored:
He hath loosed the fatal lightning of His terrible
swift sword:
 His truth is marching on.

I have seen Him in the watch-fires of a hundred
circling camps,
They have builded Him an altar in the evening
dews and damps:
I can read His righteous sentence by the dim and
flaring lamps:
 His day is marching on.

I have read a fiery gospel writ in burnished rows of
steel:
"As ye deal with my contemners, so with you my
grace shall deal;

Let the Hero, born of woman, crush the serpent
with his heel,
    Since God is marching on."

He has sounded forth the trumpet that shall never
call retreat;
He is sifting out the hearts of men before his
judgement seat;
On, be swift, my soul, to answer Him! Be jubilant,
my feet!
    While God is marching on.

In the beauty of the lilies Christ was born across the
sea,
With a glory in his bosom that transfigures you and
me;
As he died to make men holy, let us die to make
men free,
    While God is marching on.

       JULIA WARD HOWE
       "THE BATTLE HYMN OF THE REPUBLIC" (1862)

I pledge allegiance to the flag of the United States and to the republic for which it stands, one nation, indivisible, with liberty and justice for all.

FRANCIS BELLAMY
"PLEDGE OF ALLEGIANCE TO THE FLAG" (1892)

For all Americans, for all time, the ultimate question:

O, say does that star-spangled banner yet wave
O'er the land of the free and the home of the brave?

FRANCIS SCOTT KEY
"THE STAR SPANGLED BANNER" (1814)

2

Remembering Pearl Harbor

Of all the days destined to have an impact upon world history, few can equal December 7, 1941. On the United States mainland, that Sunday began with the routine innocence of church, family get-togethers, peaceful outdoor activities in the bracing late-autumn air. America was at peace, and many of her citizens were committed to the idea that her sons would never again go to fight in a foreign war, as they had for a short time in WWI.

New York City was typical of the tranquil settings that prevailed throughout America that day. At the Polo Grounds, the Giant-Dodger football game was well under way. Many fans were listening to the action on the radio, WOR. Suddenly, at 2:26, the game was interrupted for a news flash: the Japanese had attacked Pearl Harbor.

No one knew it yet, but the entire world was never going to be the same.

As the news was broadcast from the remote islands, where the attack had begun just before eight A.M. Hawaii time, the sleeping giant that was the United States of America, blinked, sat up, and stood up with clinched fists.

By evening of that day, millions of Americans would be hunched over their radios, desperate for scraps of information. Neighbor talked to neighbor, across back fences, on front porches, on street corners, in church pews. Many radios were battery-powered, particularly in southern and western rural areas where electric power was still but a promise. America's cities were there, burgeoning already, but this was still a nation of villages and hamlets and scattered farms where people were born and raised, and stayed until they died.

December 7, 1941, changed that cozy sense of isolation. From now on, Americans would go where they were needed. And from there they would go where they pleased.

Pearl Harbor wasn't just an attack, or a battle. It was a turning point in history. From the moment the first Japanese torpedo cut through the water, Japan's conquest of the Pacific region was doomed; Adolph Hitler's destruction of civilization was doomed. The price would be high. Pearl Harbor was where America began to pay that price.

Japanese future action unpredictable but hostile action possible at any moment. If hostilities cannot, repeat cannot, be avoided the United States desires that Japan commit the first overt act. This policy should not, repeat not, be construed as restricting you to a course of action that might jeopardize your defense...

WAR DEPARTMENT MESSAGE NO. 472
Signed by General George C. Marshall, Nov. 27, 1941

Climb Mount Niitaka, 1208

JAPANESE CODED ORDER, Dec. 2, 1941
Ordering Vice Admiral Chuichi Nagumo's task force to proceed with the attack on Pearl Harbor on Dec. 8 (Japan time)

This dispatch is to be considered a war warning. Negotiations with Japan looking toward stabilization of conditions in the Pacific have ceased and an aggressive move by Japan is expected within the next few days . . . Execute an appropriate defensive deployment preparatory to carrying out tasks assigned in WPL 46 [the Navy's basic war plan].

NAVY DEPARTMENT DISPATCH, Nov. 27, 1941

We have attacked, fired upon, and dropped depth charges upon submarine operating in defensive sea area.

MESSAGE FROM DESTROYER *WARD* 0653, DEC. 7, 1941
Lt. William W. Outerbridge, Captain of the destroyer *Ward*, at the channel entrance to Pearl Harbor. No action was taken on the *Ward*'s report.

Tora! Tora! Tora!
(Tiger! Tiger! Tiger!)

CODED RADIO MESSAGE 0753, DEC. 7, 1941
COMMANDER MITSUO FUCHIDA
Commander of the Japanese air strike force at Pearl
Harbor, reporting in code that total surprise had been
achieved in his attack.

Praise the Lord and pass the ammunition.

COMMANDER HOWELL FORGY
U.S. Navy Chaplain, shouting to ammo bearers on his
ship during action at Pearl Harbor, Dec. 7, 1941.

Dick, get that fellow's number, for I want to report him for about sixteen violations of the course and safety regulations.

COMMANDER LOGAN RAMSEY
0755, DEC. 7, 1941
Upon seeing a low-flying attack aircraft through the window of the Ford Island Command Center at Pearl Harbor

[Seconds later, Ramsey realized the plane in question was Japanese and raced to the radio room, ordering all radiomen to broadcast in plain English:]

AIR RAID, PEARL HARBOR. THIS IS NOT DRILL!

RADIO MESSAGE, 0758, DEC. 7, 1941

It would have been merciful had it killed me.

ADMIRAL HUSBAND KIMMEL
After picking up a spent .50-caliber machine gun bullet
that had crashed through a window where he was
standing during the Pearl Harbor attack, Dec. 7, 1941

Yesterday, December 7, 1941—a date which will
live in infamy—the United States of America was
suddenly and deliberately attacked.

PRESIDENT FRANKLIN DELANO ROOSEVELT
Addressing the Senate and Congress, Dec. 8, 1941

Right after Pearl Harbor I felt ashamed—for myself and for the entire fleet—because we hadn't been able to put up a better fight. I felt betrayed by the military brass for letting us be caught by surprise after months of preparations and training. It wasn't until after the Battle of Midway that I felt fully vindicated for the Pearl Harbor disaster. I still have ambivalent feelings about being at Pear Harbor. It wasn't a battle we won—but neither was the Alamo.

JACK WHITE
Recalling Pearl Harbor, where he was an Electrician Mate Third Class aboard the U.S.S. *New Orleans*.

Kimmel and Short were the commanders. They should have told their men to be up at the crack of dawn, scanning the skies for Japanese aircraft with their guns ready. They didn't do it . . . They were incompetent. What does it say about us that we're now rushing to exonerate Kimmel and Short? In the more than half a century since Pearl Harbor, we've developed a mentality according to which everybody has entitlements, everybody has rights, and no one ever takes responsibility for anything.

STEPHEN E. AMBROSE
WALL STREET JOURNAL, MAY 27, 1999
Upon the occasion of the U.S. Senate vote (52–47) to restore Admiral Husband Kimmel and General Walter C. Short, the commanders of the U.S. Forces in Hawaii at the time of Pearl Harbor, to their original ranks. They had been discharged in disgrace after an investigation following Pearl Harbor.

3

Under Fire

In the summer of 1999, a smallish group of Army brass, less than a squad in number, went to Hollywood. Slumming with movie stars was not their intention. They were calling on the folks who had made the film *Saving Private Ryan* and other movies to see how the Army could get some help from the filmmaking varsity. If the realism in the battle scenes in *Saving Private Ryan* could be captured in training films . . . well, it doesn't take much imagination to see how effective they could be.

Good move by the Army, I'm thinking, while hoping the project comes to fruition. Training men to function while being under fire is obviously an Army priority, and ought to be given as much realism as possible. While marches and obedience drills have always been Army training staples, combat simulations are an important objective as well. Predicting the way troops will react when real bullets are flying and shells exploding is far from an exact science, but even for an amateur like myself it is probably safe to say that the better the training, the better the soldier.

After the Korean War, the way the Army looked at the "Under Fire" subject got a jolt from Brig. Gen. S.L.A. Marshall (U.S. Army, Ret.), noted Army historian and author of several books. Marshall's *Men Against Fire* (1964) is a study of soldiers' reactions and performance when "deep in the shit," as the grunts would later say in Vietnam. One of Marshall's discoveries was the degree to which men feel separated, alone, and ineffective when out of visual contact with the individuals around them. In the smoke, noise, and confusion of the fighting, many troops would never fire their weapons. They just didn't sense the use. Today, of course, the Army and Marines try as hard as they possibly can to teach troops that one individual firing his weapon effectively amounts to a powerful fighting element. (If you want to see a movie that drives this point home relentlessly, see the late Stanley Kubrick's *Full Metal Jacket*. In it, the very lesson that was drilled into the recruits in boot camp on the effectiveness of an individual rifleman is used against them with devastating effectiveness in Vietnam.)

Apart from pragmatic purposes such as training, recollections of being under fire are the war stories we want to hear. Those of us who have never been there want to know, "What was it like?" And,

no, we don't mean the chow, or the living conditions, or being ordered around by the mean
sergeants.

"What was it like under fire?" That's the real
question.

You're about to hear some of the answers.

Under Fire would take you out of your head and
your body too . . . guys who'd played a lot of hard
sports said they'd never felt anything like it, the sud-
den drop and rocket rush of the hit, the reserves of
adrenaline . . . Unless of course you'd shit your pants
or were screaming or praying or giving anything at
all to the hundred-channel panic that blew word
salad all around you . . .

MICHAEL HERR
DISPATCHES (1968)
On the fighting in Vietnam

There was a flash, as when a blast-furnace door is swung open, and a roar . . . I tried to breathe but my breath would not come and I felt myself rush bodily out of myself . . . Then I floated, and instead of going on I felt myself slide back. I breathed and I was back.

ERNEST HEMINGWAY
A FAREWELL TO ARMS (1929)

The doggie [dogface] becomes a specialist on shells after he has been in the line awhile . . . Some shells scream, some whiz, some whistle and others whir. Most flat-trajectory shells sound like rapidly ripped canvas. Howitzer shells seem to have a two-tone whisper. Let's get the hell off this subject.

BILL MAULDIN
UP FRONT (1945)

They couldn't hit an elephant at this distance.

> UNION MAJ. GEN. JOHN SEDGWICK (1864)
> A moment later a Confederate sharpshooter's bullet
> struck him under the left eye, killing him instantly.

———

There was death all over the sky, the quiet threat of death, the anesthesia of cold sunlight filled the cockpit.

The lady named Death is a whore . . . Luck is a lady . . . and so is Death . . . And there's no telling who they'll go for. Sometimes it's a quiet, gentle, intelligent guy. The Lady Luck strings along with him for a while, and then she hands him over to the lady named Death.

> BERT STILES, WWII B-17 PILOT
> *SERENADE TO THE BIG BIRD* (1947)

They wish to hell they were someplace else, and they wish to hell they would get relief. They wish to hell the mud was dry and they wish to hell their coffee was hot. They want to go home. But they stay in their wet holes and fight, and then they climb out and crawl through minefields and fight some more.

BILL MAULDIN
UP FRONT (1945))

Dismount! Prepare to fight on foot!

CAVALRY COMMAND
Commonly given to U.S. Army Cavalry troops in the Indian campaigns of the American West

"The smoke of the shooting and the dust of the horses shut out the hill," Pte-San-Waste-Win said, "and the soldiers fired many shots, but the Sioux shot straight and the soldiers fell dead . . . Long Hair [Custer] lay dead among the rest . . . The blood of the people was hot and their hearts bad, and they took no prisoners that day."

> PTE-SAN-WASTE-WIN
> Account of the Battle of the Little Bighorn, June 25, 1876
> quoted in DEE BROWN
> *BURY MY HEART AT WOUNDED KNEE* (1970)

Benteen, come on—big village—be quick—bring packs.

GENERAL GEORGE ARMSTRONG CUSTER
Hastily scrawled message to Captain Frederick Benteen, delivered by Trumpeter Martin, the last to see Custer and his men of the 7th Cavalry alive as they rode into destiny on the Little Bighorn, June 25, 1876. ["Packs" meant pack-horses with supplies.]

Fix bayonets! Charge! Everybody goes with me!

U.S. ARMY CAPT. LEWIS MILLETT (1951)
In Korea, rallying his men for what would prove to be the most complete bayonet charge by American troops since the Civil War

War's a brain-splattering, windpipe-slitting art.

LORD BYRON (1788–1824)

Riding on the early morning mists, not knowing if the LZ is hot or cold despite endless briefings and grease pencils on acetate overlays, the constant military syntax babble between crew and pilots, the chatter of outgoing .60 fire, hot spent shells, windrush, and the knowledge that if it's a wet paddy, it's up to your thighs, chest, or above. A panorama of dragonfly-like machines stuttering into the Z, a pucker factor of f32.

TIM PAGE, COMBAT PHOTOGRAPHER
TIM PAGE'S NAM (1983)

I shan't forgit the night
When I dropped be'ind the fight
With a bullet where my belt-plate should'a'been.
I was chokin' mad with thirst,
An' the man that spied me first
Was our good old grinnin', gruntin' Gunga Din.
'E lifted up my 'ead,
And 'e plugged me where I bled,
An' 'e guv me 'arf-a-pint o' water green:
It was crawlin' and it stunk,
But of all the drinks I've drunk,
I'm gratefullest to one from Gunga Din.

RUDYARD KIPLING (1865–1936)
GUNGA DIN

Men who have been in battle know from first-hand experience that when the chips are down, a man fights to help the man next to him, just as a company fights to keep pace with its flanks. Things have to be that simple.

> BRIG. GEN. S. L. A. MARSHALL (RET.)
> *MEN AGAINST FIRE* (1964)

German prisoners, asked to assess their various enemies, have said that the British attacked singing, and the French attacked shouting, but that the Americans attacked in silence. They liked better the men who attacked singing or shouting than the grimly silent men who kept coming on stubbornly without a sound.

> JAMES JONES
> *WW II* (1975)

A speck of dirt on your windscreen could turn into an enemy fighter in the time it took to look round and back again. A little smear on your goggles might hide the plane that was coming in to kill you.

DEREK ROBINSON
PIECE OF CAKE (1983)
Describing R.A.F. action in the Battle of France and the Battle of Britain, 1940

Valor is of no service, chance rules all, and the bravest often fall by the hands of cowards.

TACITUS (A.D. 54–119)
THE HISTORIES

Come on, you sons of bitches! Do you want to live forever?

> MARINE SGT. DAN DALY (ATTRIBUTED)
> During action at Belleau Wood in WWI, June 1918

Don't fire until you see the whites of their eyes.

> WILLIAM PRESCOTT, REVOLUTIONARY WAR SOLDIER
> Command given at the Battle of Bunker Hill, 1775

A horse! A horse! My kingdom for a horse!

> KING RICHARD
> in WILLIAM SHAKESPEARE
> *KING RICHARD THE THIRD* (1593)

When your turn to watch came you lay huddled in the darkness, listening . . . straining to hear counter-attack giveaways: the whiplike crack and shrill hissing of streams of sleeting small-arms red-tracer fire, the iron ring of ricochets, and the steady belch of automatic fire. Now and then a parachute flare would burst overhead, and you could see saffron puffs of artillery in the ghostly light.

WILLIAM MANCHESTER
GOODBYE, DARKNESS (1979)
Recounting his Marine Corps combat experiences in the Pacific in WWII

An amphtrack bobbed alongside our Higgins boat. Said the Marine amphtrack boss, "Quick! Half you men get in here. They need help bad on the beach. A lot of Marines have already been killed and wounded." While the amphtrack was alongside, Jap shells from an automatic weapon began peppering the water around us . . . But the Marines did not hesitate. Hadn't they been told that other Marines "needed help bad"?

ROBERT SHERROD, WAR CORRESPONDENT
TARAWA (1944)
Describing the Marine invasion of Tarawa,
November 1943

Through the scattering, thinning mist the horizon was magically filling with ships—ships of every size and description . . . There appeared to be thousands of them. It was a ghostly armada that somehow had appeared from nowhere. At that moment the world of the good solider Pluskat began falling apart. He says that in those first few moments he knew, calmly and surely, that "this was the end for Germany."

CORNELIUS RYAN
THE LONGEST DAY (1959)
Describing the reaction of German Major Werner Pluskat at dawn on Omaha Beach, D-Day, June 6, 1944

Two kinds of people are staying on this beach, the dead and those who are going to die—now let's get the hell out of here.

COL. GEORGE A. TAYLOR, 16TH RCT, OMAHA BEACH, D-DAY, JUNE 6, 1944

As the boat rose to a sea, the green water turned white and came slamming in over the men, the guns and cases of explosives. Ahead you could see the coast of France . . . you saw the line of low, silhouetted cruisers and the two big battlewagons lying broadside to the shore. You saw the heat-bright flashes of their guns and the brown smoke that pushed out against the wind and then blew away.

ERNEST HEMINGWAY
COLLIER'S (JULY 19, 1944)
Reporting on D-day

Sighted sub, sank same.

U.S. NAVY PILOT DONALD F. MASON (1942)
One of the shortest victory messages of all time

We're the battling bastards of Bataan;
No mama, no papa, no Uncle Sam,
No aunts, no uncles, no cousins, no nieces,
No pills, no planes or artillery pieces,
And nobody gives a damn.

FRANK HEWLETT, WAR CORRESPONDENT (1942)

It was life rather than death that faded into the distance, as I grew into a state of not thinking, not feeling, not seeing. I moved past trees, past other things; men passed by me, carrying other men, some crying, some cursing, some silent. They were all unreal. Balanced uneasily on the knife-edge between utter oblivion and this temporary not-knowing, it seemed little matter whether I were . . . to go forward to death or to come back to life.

CAPTAIN WYN GRIFFITH (1916)
With the Royal Welch Fusiliers, at Mametz Wood

Give them the cold steel!

CONFEDERATE BRIG. GEN. LEWIS ARMISTEAD
Pickett's Charge at Gettysburg, July 3, 1863

There is Jackson standing like a stone wall. Let us determine to die here, and we will conquer.

BRIG. GEN. BARNARD ELLIOTT BEE
At Bull Run, July 21, 1866. Supposedly the source of
General Thomas J. Jackson's nickname, "Stonewall."

Scratch one flattop! Dixon to Carrier, Scratch one flattop!

LIEUTENANT COMMANDER R. E. DIXON
Radio message after his and other planes attacked the
Japanese carrier *Shoho*, sending it to the bottom in five
minutes during the Battle of the Coral Sea, May 7, 1942

Heaven knows its time; every bullet has its billet.

SIR WALTER SCOTT
COUNT ROBERT OF PARIS (1862)

When a fellow's time comes, down he goes. Every bullet has its billet.

CONFEDERATE PVT. E. E. PATTERSON
DIARY ENTRY (1862)

I detest the term "friendly fire." Once a bullet leaves a muzzle or a rocket leaves an airplane, it is not friendly to anyone. . . . The very chaotic nature of the battlefield, where quick decisions make the difference between life and death, has resulted in numerous incidents of troops being killed by their own fires in every war that this nation has ever fought. . . . Not even one such avoidable death should ever be considered acceptable.

GENERAL H. NORMAN SCHWARZKOPF
IT DOESN'T TAKE A HERO (1992)

We have met the enemy, and they are ours.

COMMODORE OLIVER HAZARD PERRY (1813)
In a letter to General Harrison

On account of their large draught the ships could not be breached except in deep water; and the troops, besides being ignorant of the locality, had their hands full: weighted with a mass of heavy armour, they had to jump from the ships, stand firm in the surf, and fight at the same time. But the enemy knew their ground: being quite unencumbered, they could hurl their weapons boldly from dry land or shallow water, and gallop their horses which were trained to this kind of work. Our men were terrified: they were inexperienced in this kind of fighting, and lack that dash and drive which always characterized their land battles.

JULIUS CAESAR
GALLIC WARS
Describing his first landing in Britain in 55 B.C.

The following will give you some idea of British charioteers in action. They begin by driving all over the field, hurling javelins; and the terror inspired by the horses and the noise of the wheels is usually enough to throw the enemy ranks into disorder. Then they work their way between their own cavalry units, where the warriors jump down and fight on foot. Meanwhile the drivers retire a short distance from the fighting and station the cars in such a way that their masters, if outnumbered, have an easy means of retreat to their own lines. In action, therefore, they combine the mobility of cavalry with the staying power of foot soldiers. Their skill, which is derived from ceaseless training and practice, may be judged by the fact that they can control their horses at full gallop on the steepest incline, check and turn them in a moment, run along the pole, stand on the yoke, and get back again into the chariot as quick as lightning.

> JULIUS CAESAR
> *GALLIC WARS*
> Describing his first landing in Britain in 55 B.C.

It seemed to the youth that he saw everything. Each blade of the green grass was bold and clear. He thought that he was aware of every change in the thin, transparent vapor that floated idly in sheets. The brown or gray trunks of the trees showed each roughness of their surfaces. And the men of the regiment, with their staring eyes and sweating faces, running madly, or falling, as if thrown headlong, to queer, heaped-up corpses—all were comprehended. His mind took a mechanical but firm impression, as that afterward everything was pictured and explained to him, save why he was there.

STEPHEN CRANE
THE RED BADGE OF COURAGE (1895)

The youth had a vague belief he had run miles, and he thought, in a way, that he was now in some new and unknown land.

The moment the regiment ceased its advance the protesting splutter of musketry became a steadier roar. Long and accurate fringes of smoke spread out. From the top of a small hill can level belchings of yellow flame that caused an inhuman whistling in the air.

The men, halted, had opportunity to see some of their comrades dropping with moans and shrieks. A few lay under foot, still or wailing. And now for an instant the men stood, their rifles slack in their hands, and watch the regiment dwindle. They appeared dazed and stupid. This spectacle seemed to paralyze them, overcome them with a fatal fascination. They stared woodenly at the sights, and lowering their eyes, looked from face to face. It was a strange pause, and a strange silence.

STEPHEN CRANE
THE RED BADGE OF COURAGE (1895)

He felt a quiet manhood, non-assertive but of sturdy and strong blood . . . He had been to touch the great death, and found that, after all, it was but the great death. He was a man . . . It rained. The procession of weary soldiers became a bedraggled train, despondent and muttering, marching with churning effort in a trough of liquid brown mud under a low, wretched sky. Yet the youth smiled, for he saw that the world was a world for him, though many discovered it to be made of oaths and walking sticks. He had rid himself of the red sickness of battle. The sultry nightmare was in the past. He had been an animal blistered and sweating in the heat and pain of war. He turned now with a lover's thirst to images of tranquil skies, fresh meadows, cool brooks—an existence of soft and eternal peace.

STEPHEN CRANE
THE RED BADGE OF COURAGE (1895)

On t'other side, there stood Destruction bare;
Unpunished Rapine, and a waste of war;
Contest, with sharpened knives, in cloisters drawn,
And all with blood bespread the holy lawn.
Loud menaces were heard, and foul disgrace,
And bawling infamy, in language base;
Till sense was lost in sound, and silence fled the
place.

GEOFFREY CHAUCER (1343–1400)
THE CANTERBURY TALES

I saw a large airdrome with a . . . lone jet approaching the field from the south at 500 feet. I dove at him. His landing gear was down and he was lining up the runway, coming in at no more than 200 mph, when I dropped on his ass at 500 mph . . . My hits slapped into his wings and I pulled up 300 feet off the ground with flak crackling all around me . . . I looked back and saw that jet crash-landing short of the runway, shearing off a wing, in a cloud of dust and smoke. I'd rather have brought down the son of a bitch in a dogfight . . .

GENERAL CHUCK YEAGER
YEAGER (1985)
Describing action against the German ME-262 jet fighter—the first jet plane Yeager had ever seen. He went on to become the first man to break the sound barrier.

You may fire when you are ready, Gridley.

COMMODORE GEORGE DEWEY (1898)
To the captain of his flagship at the Battle of Manilla Bay

The key to his [General Kuribayashi, Japanese Commander at Iwo Jima] battle scheme was to let the Marines land and suck them into a gigantic ambush. His troops had been stunned and shellshocked by the Americans' massive firepower, but now they were at their posts, ready to wipe out the invaders when he ordered. The shrewd samurai bided his time for nearly an hour.

BILL D. ROSS
IWO JIMA: LEGACY OF VALOR (1985)
Describing the Marine invasion of Iwo Jima,
February 19, 1945

There are moments in battles in which the soul hardens the man until the soldier is changed into a statue, and when all flesh becomes granite.

The English battalions, desperately assaulted, did not stir.

Then it was terrible.

All the faces of the English squares were attacked at once. A frenzied whirl enveloped them. That cold infantry remained impassive. The first rank knelt and received the cuirassiers on their bayonets, the second rank shot them down; behind the second rank the cannoneers charged their guns, the front of the square parted, permitted the passage of an eruption of grape-shot, and closed again. The cuirassiers replied by crushing them. Their great horses reared, strode across the ranks, leaped over the bayonets and fell, gigantic, in the midst of these four living walls. The cannon-balls ploughed furrows in these cuirassiers; the cuirassiers made breaches in the squares. Files of men disappeared, ground to dust under the horses.

VICTOR HUGO
LES MISÉRABLES (1862)
Describing the Battle of Waterloo, June 1815

Up, Guards, and at them!

THE DUKE OF WELLINGTON
Rallying his troops at the Battle of Waterloo, June 1815

The bayonets plunged into the bellies of these centaurs; hence a hideousness of wounds which has probably never been seen anywhere else. The squares, wasted by this mad cavalry, closed up their ranks without flinching. Inexhaustible in the matter of grape-shot, they created explosions in their assailants' midst. The form of this combat was monstrous. There squares were no longer battalions, they were craters; those cuirassiers were no longer cavalry, they were a tempest. Each square was a volcano attacked by a cloud; lava combated with lightning.

VICTOR HUGO
LES MISÉRABLES (1862)
Describing the Battle of Waterloo, June 1815

The red regiment of English Guards, lying flat behind the hedges, sprang up, a cloud of grape-shot riddled the tricoloured flag and whistled round our eagles; all hurled themselves forwards, and the supreme carnage began. In the darkness, the Imperial Guard felt the Army losing ground around it, and in the vast shock of the rout it heard the desperate flight which had taken the place of "Long live the Emperor!" and, with flight behind it, it continued to advance, more crushed, losing more men at every step it took. There were none who hesitated, no timid men in its ranks. The soldier in that troop was as much of a hero as the general. Not a man was missing in that heroic suicide.

VICTOR HUGO
LES MISÉRABLES (1862)
Describing the Battle of Waterloo, June 1815

But now from the direction of the enemy there came a succession of grisly apparitions; horses spouting blood, struggling on three legs, men staggering on foot, men bleeding from terrible wounds, fish-hook spears stuck right through them, arms and faces cut to pieces, bowels protruding, men gasping, crying, collapsing, expiring.

WINSTON CHURCHILL
A ROVING COMMISSION, "THE CAVALRY CHARGE AT OMDURMAN" (1930)

Lovely weather for killing Germans.

GENERAL GEORGE S. PATTON, JR.
On seeing the weather breaking over the Bulge and knowing his counter-attacking forces would now have air support, Christmas morning, 1944

Almighty and most merciful Father, we humbly beseech Thee, of Thy great goodness, to restrain these immoderate rains with which we have had to contend. Grant us fair weather for battle. Graciously hearken to us as soldiers who call upon Thee that armed with Thy power, we may advanced from victory to victory, and crush the oppression and wickedness of our enemies, and establish Thy justice among men and nations. Amen.

CHAPLAIN JAMES H. O'NEILL
Chaplain of Third Army, O'Neill wrote the prayer at the order of General George Patton during the Battle of the Bulge, December 1944.

I love combat. I hate war. I don't understand it, but that's the way it is.

> GENERAL CHUCK HORNER
> Air Force pilot in Vietnam and later Air Force
> Commander during the Gulf War
> *EVERY MAN A TIGER* (1999)

I gave him the whole nine yards!

> P-51 MUSTANG PILOT
> WWII expression referring to the plane's 27-foot-long
> ammo belt.

4

Uncommon Valor

"I was the homesickest boy you ever saw."

The "boy" who wrote those words in his diary at Camp Gordon, Georgia, in 1917, was a husky, red-headed six-footer almost 30 years old who was destined to become a living legend.

When he left Pall Mall, Tennessee, in the remote mountain valley called the Three Forks of the Wolf, Alvin York had already fought and won a battle many men could not endure. One of eleven children raised in mountain ways, with scant "book-l'arnin"— mainly the three Rs—York was taught to believe the words of the New Testament that stated, "For all they that take the sword shall perish with the sword." So involved with his church was York that he had clear grounds for an exemption from military service. He would have none of it, for though he believed it was wrong to kill other men, he believed it necessary to serve his country. To the dismay of all around him, and despite being engaged to be married, the experienced hunter and marksman had set off for Camp Gordon.

Less than a year later, Cpl. Alvin York would kill 20 German soldiers and capture 132—plus a battalion commander and 35 guns—virtually single-handedly in a single action. He would be awarded the Congressional Medal and, as Sergeant York, become one of America's most celebrated and legendary war heroes.

Where do we get heroes such as Sergeant York? Are they simply men who know no fear?

No way, the fiery General George S. Patton Jr. tells us. "No sane man is unafraid in battle, but discipline produces in him a form of vicarious courage," Patton says in *War As I Knew It* (1941).

More recently, General H. Norman Schwarzkopf sees heroism in every soldier who goes into combat: "It doesn't take a hero to order men into battle. It takes a hero to be one of those men who goes into battle."

This section is about courage. It would seem that valor in combat—and courage "above and beyond the call of duty," as in the Medal of Honor—cannot be predicted, commanded, or analyzed. But it can damn sure be remembered.

Uncommon valor was a common virtue.

ADMIRAL CHESTER NIMITZ
SPECIAL COMMUNIQUE, MARCH 17, 1945
Declaring the island of Iwo Jima to be "officially
secured," and saluting the bravery of the Americans who
had fought there

It doesn't take a hero to order men into battle. It
takes a hero to be one of those men who goes into
battle.

GEN. H. NORMAN SCHWARZKOPF
IT DOESN'T TAKE A HERO (1992)

By the rude bridge that arched the flood,
Their flag to April's breeze unfurled,
Here once the embattled farmers stood
And fired the shot heard round the world.

> RALPH WALDO EMERSON
> "CONCORD HYMN" (1837)
> Sung at the completion of the Battle of Concord
> monument on July 4, 1837

When valour preys on reason,
It eats the sword it fights with.

> WILLIAM SHAKESPEARE
> *ANTONY AND CLEOPATRA (1607)*

A hero is no braver than an ordinary man, but he is brave five minutes longer.

> RALPH WALDO EMERSON (1803–1882)

War is fear cloaked in courage.

GENERAL WILLIAM WESTMORELAND
MCCALLS (1966)

"Forward the Light Brigade!"
Was there a man dismayed?
Not though the soldier knew
Someone had blundered.
Theirs not to make reply,
Theirs not to reason why.
Theirs but to do and die.
Into the valley of Death
       Rode the six hundred.

Cannon to the right of them,
Cannon to the left of them,
Cannon in front of them
Volleyed and thundered;

Stormed at with shot and shell,
Boldly they rode and well,
Into the jaws of Death,
Into the mouth of hell
     Rode the six hundred.

ALFRED LORD TENNYSON
"THE CHARGE OF THE LIGHT BRIGADE" (1854)

We gained nothing but glory and lost our bravest men.

CONFEDERATE OFFICER
quoted in SHELBY FOOTE , THE CIVIL WAR: A NARRATIVE (1963)
On the failure of Pickett's Charge at Gettysburg,
July 3, 1883

TO THE U.S.A. COMMANDER OF THE ENCIR-
CLED TOWN OF BASTOGNE:

THE FORTUNE OF WAR IS CHANGING. THIS
TIME, THE U.S.A. FORCE IN AND NEAR BAS-
TOGNE HAVE BEEN ENCIRCLED BY STRONG
GERMAN ARMORED UNITS . . . THERE IS ONLY
ONE POSSIBILITY OF SAVING THE ENCIRCLED
U.S.A. TROOPS FROM TOTAL ANNIHILATION:
THE HONORABLE SURRENDER OF THE ENCIR-
CLED TOWN. IN ORDER TO THINK IT OVER,
A TERM OF TWO HOURS WILL BE GRANTED BE-
GINNING WITH THE DELIVERY OF THIS NOTE . .
. THE ORDER FOR FIRING WILL BE GIVEN IM-
MEDIATELY AFTER THIS TWO HOURS' TERM.

ALL THE SERIOUS CIVILIAN LOSSES CAUSED BY THIS ARTILLERY FIRE WOULD NOT CORRESPOND WITH THE WELL-KNOWN AMERICAN HUMANITY.

THE GERMAN COMMANDER quoted in HANSON W. BALDWIN, *THE BATTLE OF THE BULGE* (1966)
Message from the Germans delivered under a flag of truce to the 101st Airborne, December 1944

To the German Commander: NUTS! The American Commander.

BRIG. GEN. ANTHONY MCAULIFFE quoted in STEPHEN E. AMBROSE, *CITIZEN SOLDIERS* (1997)
The 101st Airborne Commander's reply to German demand for surrender while surrounded at Bastogne, Battle of the Bulge, December 1944

I only regret that I have but one life to lose for my country.

NATHAN HALE
Before being executed as a spy by the British, 1776

———•••———

Heroism feels and never reasons, and therefore is always right.

RALPH WALDO EMERSON
"HEROISM" (1841)

The high sentiments always win in the end, the leaders who offer blood, toil, tears and sweat always get more out of their followers than those who offer safety and a good time. When it comes to the pinch, human beings are heroic.

GEORGE ORWELL
"THE ART OF DONALD MCGILL,"
HORIZON (Sept., 1941)

Courage is doing what you're afraid to do. There can be no courage unless you're scared.

CAPT. EDDIE RICKENBACKER, WWI AMERICAN AIR ACE
quoted in *THE NEW YORK TIMES* (1963)

For those who were there, or whose friends were there, Guadalcanal is not a name but an emotion, recalling desperate fights in the air, furious night naval battles, frantic work at supply or construction, savage fighting in the sodden jungle, nights broken by screaming bombs and deafening explosions of naval shells. Sometimes I dream of a great battle monument on Guadalcanal; a granite monolith on which the names of all who fell and of all ships that rest in Ironbottom Sound may be carved. At other times I feel that the jagged cone of Savo Island, forever brooding over the blood-thickened waters of the Sound, is the best monument to the men and ships who here rolled back the enemy tide.

SAMUEL ELIOT MORISON
THE STRUGGLE FOR GUADALCANAL, VOLUME FIVE, HISTORY OF UNITED STATES NAVAL OPERATIONS IN WORLD WAR TWO (1949)

Cowards die many times before their deaths;
The valiant never taste of death but once.
Of all the wonders that I yet have heard,
It seems to me most strange that men should fear;
Seeing that death, a necessary end,
Will come when it will come.

> JULIUS CAESAR
> in WILLIAM SHAKESPEARE
> *JULIUS CAESAR* (1599)

O God of battles! Steel my soldiers' hearts;
Possess them not with fear; take from them now
The sense of reckoning, if the opposed numbers
Pluck their hearts from them.

> KING HENRY
> in WILLIAM SHAKESPEARE
> *KING HENRY THE FIFTH* (1589)

The black rank and file volunteered when disaster clouded the Union cause, served without pay for eighteen months till given that of white troops, faced threatened enslavement if captured, were brave in action, patient under heavy and dangerous labors, and cheerful amid hardships and privations.

Together they gave to the nation and the world undying proof that Americans of African descent possess the pride, courage, and devotion of the patriot soldier. One hundred and eighty thousand such Americans enlisted under the Union flag in 1863–1865.

INSCRIPTION, ROBERT GOULD SHAW
MONUMENT, BOSTON COMMONS (1897)

I have not yet begun to fight.

> CAPT. JOHN PAUL JONES
> During the battle between the *Bonhomme Richard* (Jones'
> ship) and the *Serapis*, Sept. 23, 1779

Damn the torpedoes! Captain Drayton, go ahead!
Jouett, full speed!

> ADMIRAL DAVID GLASGOW FARRAGUT
> MOBILE BAY, 1864
> The "torpedoes" he describes would be described today
> as "tethered mines."

Of all the Spartans and Thespians who fought so valiantly on that day [at Thermopylae, 480 BC], the most signal proof of courage was given by the Spartan Dieneces. It is said that before the battle he was told by a native of Trachis that, when the Persians shot their arrows, there were so many of them that they hid the sun. Dieneces, however, quite unmoved by the thought of the terrible strength of the Persian army, merely remarked: "This is pleasant news that the stranger from Trachis brings us: for if the Persians hide the sun, we shall have our battle in the shade."

Herodotus (484–424 B.C.)
THE HISTORIES

Bayonet!

COLONEL JOSHUA CHAMBERLAIN
quoted in HARRY W. PFANZ,
GETTYSBURG, THE SECOND DAY (1987)
Ordering the 20th Maine to charge attacking
Confederates at Little Round Top, Gettysburg, July 2, 1863

All right, men, we can die but once. This is the time and place. Let us charge.

UNION BRIG. GEN. WILLIAM HAINES LYTLE
Just before his death at the Battle of Chickamauga, 1863

"Holy Father Tiber, I pray thee, receive these arms, and this thy soldier, in thy propitious stream."

HORATIUS COCLES, ROMAN SOLDIER
quoted in LIVY (59 B.C.–A.D. 17)
Leaping into the Tiber after refusing to yield a bridge to
overwhelming forces

One of our platoon sergeants, during a relatively light Japanese attack on his position, reached into his hip pocket for a grenade he'd stuck there, and got it by the pin. The pin came out but the grenade didn't. No one really knows what he thought about during those split seconds. What he did was to turn away and put his back against a bank to smother the grenade away from the rest of the men. He lived maybe five or ten minutes afterward, and the only thing he said, in a kind of awed, scared, very disgusted voice, was, "What a fucking recruit trick to pull."

James Jones
WW II (1975)
Recalling his Army experiences fighting on Guadalcanal

Ney [French Marshal Michel Ney], bewildered, great with all the grandeur of accepted death, offered himself to all blows in that tempest. He had his fifth horse killed under him there. Perspiring, his eyes aflame, foam on his lips, with uniform unbuttoned, one of his epaulets half cut off by a sword-stroke from the horse-guard, his plaque with the great eagle dented by a bullet; bleeding, bemired, magnificent, a broken sword in his hand, he said, "Come and see how a Marshal of France dies on the field of battle!" But in vain; he did not die.

VICTOR HUGO
LES MISÉRABLES (1862)
Describing the Battle of Waterloo, June 1815

So there is nothing for me! Oh! I should like to have all these English bullets enter my chest!

> MARSHAL MICHEL NEY
> in VICTOR HUGO
> *LES MISÉRABLES,* (1862)
> In frustration and anger, seeing the Battle of Waterloo about to be lost

That officer is General Frazer; I admire him, but he must die. Our victory depends on it. Take your stations in that clump of bushes and do your duty.

> AMERICAN COL. MORGAN
> Order to his men at the Battle of Saratoga after observing Brig. Gen. Frazer rallying his British troops (1777)

They were expendable.

> WILLIAM L. WHITE
> Title of White's book about U.S. Torpedo Boats in the early fighting in the Pacific in WWII.

5

At Ease!

Among the countless contradictions of military life is the fact that while it might not often be *fun*, there are many things about it that are very, very *funny*.

These days, what once was an entertainment staple—service comedy—has fallen into short supply, swept away in the fallout of Vietnam, no doubt.

But while it may seem, at this point in time, that we have heard the inevitable "Taps" for service comedy, I urge you to fear not.

Even though the time for "Lights Out" in the barracks has seemingly passed, somewhere out there in the dark a flashlight glows beneath a GI blanket, and creative mischief is being plotted by the next *Sergeant Bilkos*, the bad boys and girls of *M*A*S*H*, the stalwart *Mister Roberts*, the swabies of *McHale's Navy*, the unflappable Andy Griffith in *No Time for Sergeants*, the wayward cavalrymen of *F-Troop*. Perhaps the master of comedy, Neil Simon, can come up with something new. His *Biloxi Blues* wasn't exactly chopped liver.

The laughter will return. Count on it.

In the meantime, I offer this sampler of some of service mirth's most memorable moments.

There was only one catch and that was Catch 22.Orr would be crazy to fly more missions and sane if he didn't, but if he was sane he had to fly them. If he flew them he was crazy and didn't have to; but if he didn't want to he was sane and had to.

JOSEPH HELLER
CATCH 22 (1961)

———•••••———

Yea, though I walk through the valley of the shadow of death, I will fear no evil. For I am the strongest, meanest, most-heavily-armed motherfucker in the whole damn valley!

ANONYMOUS (C. VIETNAM WAR)

Nothing in life is so exhilarating as to be shot at without result.

WINSTON CHURCHILL
THE STORY OF THE MALAKAND FIELD FORCE (1898)

Got paid out on Monday
Not a dog soldier no more
They gimme all that money
So much my pockets is sore
More dough than I can use. Re-enlistment Blues.

JAMES JONES
"THE RE-ENLISTMENT BLUES,"
FROM HERE TO ETERNITY (1951)

I can't wake'em up, I can't wake'em up, I can't
wake'em up in the morning,
I can't wake'em up, I can't wake 'em up, I can't
wake 'em up at all.
The corporal's worse than the private, the
sergeant's
worse than the corporal,
The lieutenant's worse than the sergeant, and the
captain's worse of all.

"REVEILLE" BUGLE CALL (ONE VERSION OF MANY)

When the military man approaches, the world locks
up its spoons and packs off its womankind.

GEORGE BERNARD SHAW (1856–1950)

It was so wonderful, Joe. You never heard such cheering.

> MARILYN MONROE
> quoted in GAY TALESE, *PORTRAITS* (1970)
> Speaking of entertaining troops in Korea (1954)

Yes I have.

> JOE DIMAGGIO
> quoted in GAY TALESE, *PORTRAITS* (1970)
> Answering his wife

A prisoner of war is a man who tries to kill you and fails, and then asks you not to kill him.

> WINSTON CHURCHILL (1874–1965)

Run, old hare. If I was a old hare, I'd run too.

> CONFEDERATE SOLDIER
> quoted in SHELBY FOOTE, *THE CIVIL WAR* (1963)
> Upon seeing a fleeing rabbit at the beginning of Pickett's
> Charge, Gettysburg, July 3, 1863

G.I. Joe.

> CARTOONIST DAVID BREGER
> For a WWII comic strip in *Yank*. The name is based on
> the Army term "government issue."

. . . in response to complaints that it took me five times longer to write the war than the participants took to fight it, I would point out that there were a good many more of them than there was of me.

> SHELBY FOOTE
> *THE CIVIL WAR* (1974)
> In "Note" ending his monumental three-volume classic
> on the Civil War

If it moves, salute it; if it doesn't move, pick it up; and if you can't pick it up, paint it.

ANONYMOUS (C. WWII)

True pacifism is the finest form of manliness. But if a man comes up to you and cuts your hand off, you don't offer him the other one. Not if you want to go on playing the piano, you don't.

SAM PECKINPAH, FILM DIRECTOR
PLAYBOY (1972)

"Peace" is where nobody's shooting. A "Just Peace" is when our side gets what it wants.

BILL MAULDIN, SOLDIER AND CARTOONIST
quoted in *LOOSE TALK*, LINDA BOTTS, ED. (1980)

Piss when you can.

ATTRIBUTED TO THE DUKE OF WELLINGTON,
ARTHUR WELLESLEY (1769–1852)
When the leader of the Battle of Waterloo was asked by a
young officer for his best piece of permanent military
wisdom

Only the winners decide what were war crimes.

GARY WILLS
THE NEW YORK TIMES (1975)

Out here [Vietnam] due process is a bullet.

JOHN WAYNE
in the film *THE GREEN BERETS* (1968)
SCREENPLAY, JAMES LEE BARRETT, BASED ON THE NOVEL BY
ROBIN MOORE

Captain, this is Ensign Pulver. I just threw your palm trees overboard. Now what's all this crap about no movie tonight?

ENSIGN PULVER
in *MISTER ROBERTS,* PLAY BY THOMAS HEGGEN AND
JOSHUA LOGAN (1948)
Fictional character on board a supply ship in the
Pacific in WWII

Tell me what brand of whiskey that Grant drinks. I would like to send a barrel of it to my other generals.

ABRAHAM LINCOLN (APOCRYPHAL)

"Men," a sergeant told his people aboard ship before our invasion of the island, "Saipan is covered with dense jungle, quicksand, steep hills and cliffs hiding batteries of huge coastal guns, and strongholds of reinforced concrete. Insects bear lethal poisons. Crocodiles and snakes infest the streams. The waters around it are thick with sharks. The population will be hostile toward us." There was a long silence. Then a corporal said, "Sarge, why don't we just let the Japs keep it?"

WILLIAM MANCHESTER
GOODBYE, DARKNESS (1979)

It's a long way to Tipperary, it's a long
way to go;
It's a long way to Tipperary, to the
        sweetest girl I know!
Good-bye, Piccadilly, farewell,
        Leicester Square.
It's a long, long way to Tipperary, but
        my heart's right there!

"TIPPERARY" (1908)
SONG BY HARRY WILLIAMS AND JACK JUDGE
Very popular with British soldiers in WW I. Tipperary is a
town in Ireland.

—————

When the going gets tough, they call out the sons of
bitches.

FLEET ADMIRAL ERNEST J. KING (C. WWII)

The answer is "Chicken Teriyaki." What is the question?

> *JOHNNY CARSON SHOW* (C. 1980s)
> The question is: "What is the name of the oldest living kamikaze pilot?"

The eagle craps today.

> ANONYMOUS
> Old Army expression for payday

"Close" only counts in horseshoes and hand grenades.

> ANONYMOUS

[The Emperor Vespasian (A.D. 69–79)] missed no opportunity tightening discipline; when a young man reeking of perfume came to thank him for a commission he had asked for and obtained, Vespasian turned his head away in disgust and cancelled the order, saying crushingly: "I should not have minded so much if it had been garlic."

SUETONIUS
LIVES OF THE TWELVE CAESARS,
ROBERT GRAVES, TRANS.

This durn fight ain't got any rear.

WOUNDED SOLDIER (SHILOH, 1862)
quoted in *CIVIL WAR QUOTATIONS,*
DARRYL LYMAN, ED. (1995)
Comment when ordered to the rear by his captain at Hornets' Nest

The hen is the wisest of all the animal creation because she never cackles until after the egg is laid.

> ABRAHAM LINCOLN
> Commenting on Union General Joe Hooker's incessant boasting, 1863

A crow could not fly over it without carrying his rations with him.

> UNION GEN. PHILIP HENRY SHERIDAN
> Boasting after his Army's destruction of the Shenandoah Valley of Virginia, 1864

The race is not always to the swift nor the battle to the strong, but that's the way to bet.

> ATTRIBUTED TO DAMON RUNYON

It was easy. They sank my boat.

PRESIDENT JOHN F. KENNEDY
A THOUSAND DAYS (1965)
Reply when asked how he became a hero

Montgomerys.

ERNEST HEMINGWAY
quoted in MICHAEL REYNOLDS, *HEMINGWAY:
THE FINAL YEARS* (1999)
His expression for a shaker of martinis made with a 15 to
1 gin-vermouth mixture, sarcastically referring to British
Field Marshal Bernard Law Montgomery's incessant
striving for a 15 to 1 advantage on the battlefield

Some people live an entire lifetime and wonder if they ever made a difference in the world, but the Marines don't have that problem.

RONALD REAGAN (C. 1980s)

They're over-paid, over-sexed and over here.

ANONYMOUS
British quip about American servicemen stationed in England during the buildup for D-Day, 1944

6

Soldiering Through

I'm a professional soldier, born across the road from Wellington Barracks. I enlisted when I was 17. By the time I was 24, I was a sergeant, serving on the northwest frontier of India. Sometimes . . . I would lie awake in my tent with a hurricane lamp, sometimes in the middle of a blizzard, reading about the exploits of other British soldiers. Sometimes I'd be lying there in my freezing cold tent, actually sweating. Beads of sweat pouring down my face from a battle 200 years old . . .

RICHARD ATTENBOROUGH, AS SERGEANT MAJOR
LAUDERDALE, IN THE FILM *GUNS AT BATASI* (1964)

For all the personal satisfactions and personal achievements that are possible in military life, there exists a thorny undergrowth of difficulties that every man and woman in uniform must conquer, one way or another. These obstacles cannot be ignored, out-flanked, or smashed by sheer physical strength. Their names are Chicken Shit, Boredom, and Fatigue.

Chicken Shit: mindless, unnecessary tasks invented by incomptent or demented non-coms and officers in the name of prepardnesses or discipline. Boredom: endless waiting, delays, stalls, time when you are neither *here* nor *there*, glacier-slow passage of time without challenge or reward. Fatigue: mind-numbing cleaning, fixing, packing, unpacking, lifting and toting, marching, humping.

Beyond all this come the drills and ceremonies, and somewhere along the way one must find a way to eat and sleep—and all the while be engaged in training that will hopefully equip you to do your job and stay alive when the bad people are shooting at you. If your job happens to be in the combat arms, or takes you to places like Vietnam where the entire theater of operations is virtually a combat zone, you might even see action. Get "into the shit" as the grunts said in Vietnam.

If the moments of drudgery in military life could smother the ideals and dedication that make outstanding men and women want to serve their country, America would have been put out of business long ago. Instead, we have been more than blessed with the kind of individuals who, having taken the oath, find a way to press on through whatever difficulties stand between them and doing their jobs.

American servicemen and women, it would seem from the evidence, are not easily stopped by friendly fire or enemy fire; by red tape and bureaucratic wrangling; by fools and horse's asses who sometimes wiggle their way into command; by hard work, bad weather, or poor or no food.

Soldiering through! It means hanging in there— no matter what. Very few people in civilian life have to be so committed.

The first qualification of a soldier is fortitude under fatigue and privation. Courage is only the second; hardship, poverty, and want are the best school for a soldier.

NAPOLEON I
MAXIMS, LVIII (1831)

A pint of sweat will save a gallon of blood.

GEN. GEORGE S. PATTON, JR.
In a message to troops before "Operation Torch," 1942

My business is stanching blood and feeding fainting men; my post the open field between the bullet and the hospital.

CLARA BARTON
UNION NURSE AND FOUNDER OF THE AMERICAN RED CROSS,
In a letter, 1863

An army is a team, lives, sleeps, fights, and eats as a team. This individual hero stuff is a lot of horseshit.

GEN. GEORGE S. PATTON, JR.
To troops during D-Day buildup in England, 1944

There can be no retreat when there's no rear. You can't retreat, or even withdraw, when you're surrounded. The only thing you can do is to break out, and in order to do that you have to attack, and that is what we're about to do. Heck, all we're doing is attacking in a different direction.

> MARINE MAJ. GEN. OLIVER P. SMITH
> quoted in MARTIN RUSS, *BREAKOUT* (1999)
> Speaking to a British reporter at Hagaru, North Korea, after invading Chinese troops had cut off Marine and Army forces at the Chosin Reservoir, December 1950

Retreat, hell. We just got here!

> MARINE CAPT. LLOYD WILLIAMS (1918)
> quoted in MARTIN RUSS, *BREAKOUT* (1999)
> Answering a messenger from the French Commander as Marines arrive at the Belleau Wood Sector in WWI.

The marines have a way of making you afraid, not of dying, but of not doing your job.

CAPT. BONNIE LITTLE
quoted in BILL ROSS, *IWO JIMA: LEGACY OF VALOR* (1985)

Our brains ache, in the merciless iced east winds that knife
us . . .
Wearied we keep awake because the night is silent . . .
Low, drooping flares confuse our memory of the salient . . .
Worried by silence, sentries whisper, curious, nervous.
But nothing happens.

WILFRED OWEN (1893–1918)
"EXPOSURE"

Upon the fields of friendly strife
Are sown the seeds
That upon other fields, on other days
Will bear the fruits of victory.

> GENERAL DOUGLAS MACARTHUR
> Words he wrote to hang over the portals of the
> gymnasium of the U.S. Military Academy, West Point,
> while he was serving as Superintendent, 1919–1922.

There were other opportunities, making a lot more money than I am now. Every time I have faced up to this choice, I just find the satisfaction of being a soldier and the love of my profession overwhelming and more important to me than making a great deal of money or doing something I may not like as much as being a solider.

> GENERAL COLIN POWELL
> *IN HIS OWN WORDS: COLIN POWELL,*
> LISA SHAW, ED. (1995)

The ideal officer in any army knows his business. He is firm and just ... An officer is not supposed to sleep until his men are bedded down. He is not supposed to eat until he has arranged for his men to eat. He's like a prizefighter's manager. If he keeps his fighter in shape the fighter will make him successful. I respect those combat officers who feel this responsibility so strongly that many of them are killed fulfilling it.

BILL MAULDIN
UP FRONT (1945)

Women in uniform were a new phenomenon for the Americans of the 1940s. They were in every branch of service, but more strictly segregated by their sex than blacks were by their race . . . The women in uniform did everything the men did, except engage in combat . . . Eisenhower felt he could not have won the war without them.

STEPHEN E. AMBROSE
D-DAY (1994)

If I had learned to type, I never would have made brigadier general.

BRIG. GEN. ELIZABETH P. HOSINGTON
quoted in *THE NEW YORK TIMES*,
JUNE 20, 1970

There is many a boy here today who looks on war as all glory, but, boys, it is all hell.

UNION GENERAL WILLIAM TECUMSEH SHERMAN
Speech at Columbus, Ohio, 1880

I've been aboard this destroyer for two weeks now . . .
I'm in the war at last, Doc . . . But I'm thinking now of
you, Doc . . . and everybody else aboard that bucket—
all the guys everywhere who sail from Tedium to Ap-
athy and back again—with an occasional side trip to
Monotony . . . I've discovered, Doc, that the most terrible enemy of the war is the boredom that eventually
becomes a faith and, therefore, a sort of suicide—and
I know now that the ones who refuse to surrender to
it are the strongest of all.

MISTER ROBERTS
in *MISTER ROBERTS*, PLAY BY THOMAS HEGGEN AND JOSHUA
LOGAN (1948). Fictional character's letter to his former
ship in the Pacific, 1945. Portrayed by Henry Fonda in the
original Broadway production and later in the film.

I had been living for a long time with the knowledge that if I ever shut my eyes in the dark and let myself go, ever since I had been blown up at night and felt it go out of me and go off and then come back, I tried never to think about it, but it had started to go since, in the nights, just at the moment of going off to sleep, and I could only stop it by a very great effort. So while now I am fairly sure that it would not really have gone out, yet then, that summer, I was unwilling to make the experiment.

ERNEST HEMINGWAY
"NOW I LAY ME," *MEN WITHOUT WOMEN* (1927)
Short story drawing upon his experience of being wounded as a Red Cross ambulance driver in Italy, 1918.

Give them great meals of beef and iron and steel, they will eat like wolves and fight like devils.

> CONSTABLE
> in WILLIAM SHAKESPEARE
> *KING HENRY THE FIFTH* (1599)

It was almost noon when they crossed the Korean coast. Cleve stared anxiously at it, drifting past beneath the wing. . . . He began to experience something of the exhilaration that came with triumph. In this war, he was more certain than ever, he would attain himself, as men do who venture past all that is known.

> CAPTAIN CLEVE CONNELL
> in JAMES SALTER, *THE HUNTERS* (1956)
> Fictional character on his way to fly F-86 fighters against Soviet MIGs in the Korean War.

The 1000-Yard Stare

COMMON EXPRESSION
Used to describe a sort of "vacant glare" of eyes that
have seen too much combat and now mirror no feeling
or recognition

Shall I ask the brave soldier, who fights by my side
In the cause of mankind, if our creeds agree?
Shall I give up the friend I have valued and tried,
If he kneel not before the same altar with me?

THOMAS MOORE (1780–1852)
"COME, SEND ROUND THE WINE"

"What do you want to go back to the Army for?" she cried, getting her breath. "What did the Army ever do for you? Besides beat you up and treat you like scum, and throw you in jail like a criminal? What do you want to go back to that for?"

"What do I want to go back for?" Prewitt said wonderingly. "I'm a soldier."

JAMES JONES
FROM HERE TO ETERNITY (1951)
In Jones' novel, the AWOL Private Prewitt is killed by soldiers on guard as he attempts to return to his unit following the Pearl Harbor attack.

The infantry doesn't change. We're the only arm [of the military] where the weapon is the man himself.

MAJ. GEN. C. T. SHORTS, BRITISH DIRECTOR OF INFANTRY
quoted in *NEW YORK HERALD TRIBUNE*, FEB. 4, 1985

Soldiers ought more to fear their general than their enemy.

MICHEL DE MONTAIGNE (1533–1592)
ESSAYS, BOOK III

F.U.B.A.R.

WWII EXPRESSION RE-POPULARIZED BY THE FILM
SAVING PRIVATE RYAN BY ROBERT RODAT, DIR. BY
STEVEN SPIELBERG (1998)
Acronym for "Fucked Up Beyond Any Remedy" or
sometimes as "... Beyond All Recognition."

He did not want any consequences. He did not want any consequences ever again. He wanted to live along without consquences. Besides he did not really need a girl. The army had taught him that . . . You did not need a girl unless you thought about them . . . Then sooner or later you always got one. When you were really ripe for a girl you always got one. You did not have to think about it. Sooner or later it would come. He had learned that in the army.

> HAROLD KREBS
> in ERNEST HEMINGWAY, "SOLDIER'S HOME,"
> *IN OUR TIME* (1925)
> Ernest Hemingway's fictional soldier, home from WWI

The iron-armed soldier, the true-hearted soldier,
The gallant old soldier of Tippecanoe.

> CAMPAIGN SONG FOR WILLIAM HENRY HARRISON (1840)

This is my rifle.

This (pointing to male sex organ) is my gun.

This is for fighting.

This (pointing) is for fun.

ARMY AND MARINE BOOT CAMP EXPRESSION
Popular during WWII to teach recruits to use the word
"rifle" instead of "gun" when referring to their weapon

On Parris Island . . . you were told that there were three ways of doing things: the right way, the wrong way, and the Marine Corps way. The Corps way was uncompromising.

WILLIAM MANCHESTER
GOODBYE, DARKNESS (1979)
Recalling his Marine service during WWII

Kilroy was here.

ANONYMOUS
Famous expression that popped up as grafitti during
WWII. The author was never positively identified.

Semper fidelis
Ever faithful

MOTTO OF THE U.S. MARINE CORPS

Semper Fi, Mac!

COMMON EXPRESSION, MARINE TO MARINE

Semper paratus
Always ready

U.S. COAST GUARD MOTTO (1915)

The outfit soon took on color, dash and a unique flavor which is the essence of that elusive and deathless thing call soldiering.

GENERAL DOUGLAS MACARTHUR
quoted in *LIFE* magazine, Jan. 24, 1964, on the
"Rainbow Division" in WWI

No one remembered separate days anymore, and history, being made each day, was never noticed but only merged into a great blur of tiredness and dust, of the smell of dead cattle, the smell of earth new-broken by TNT, the grinding sound of tanks and bulldozers, the sound of automatic-rifle and machine-gun fire, the interceptive dry rattle of German machine-pistol fire, dry as a rattler rattling; and the quick, spurting tap of the German light machine guns . . .

ERNEST HEMINGWAY, *COLLIER'S* magazine, Nov. 4, 1944
Reporting on the fighting after D-Day

There's old "blood and guts."

GI EXPRESSION FOR GENERAL GEORGE S. PATTON, JR. THIRD ARMY, WWII

You'll be taught food drill, the handling of a rifle, the use of the gas mask, the peculiarities of military vehicles, and the intricacies of military courtesy . . . You'll be initiated into the mysteries of the kitchen police . . . All your persecution is deliberate, calculated, systematic . . . applied to the grim and highly important task of transforming a civilian into a soldier, a boy into a man. It is the Hardening Process.

PVT. MARION HARGROVE
SEE HERE, PRIVATE HARGROVE (1942)

The future is not always easy to predict and our record regarding where we will fight future wars is not the best. If someone had asked me on the day I graduated from West Point, in June 1956, where I would fight for my country during my years of service, I'm not sure what I would have said. But I'm damn sure I would *not* have said, "Vietnam, Grenada, and Iraq."

GEN. H. NORMAN SCHWARZKOPF
IT DOESN'T TAKE A HERO (1992)

There is always inequity in life. Some men are killed in a war, and some men are wounded, and some men are stationed in the Antarctic and some are stationed in San Francisco. It's very hard in military or personal life to assure complete equality. Life is unfair.

PRESIDENT JOHN F. KENNEDY
Press conference, speaking of calling up
reservists for the "Berlin Wall Crisis" (1962)

There *was* an afternoon for one Marine, and he spent it carefully learning to light and keep alive the warmth of a peasant's charcoal brazier, upon which he heated a single can of beans and a canteen cup of coffee. And that was victory. The freedom to sprawl loosely upon a city street, heat his coffee and eat a can of beans . . . with no enemy bullets forcing him to toss the can aside while diving behind another wall for momentary survival.

DAVID DOUGLAS DUNCAN
THIS IS WAR (1951)
Describing the war in Korea during the winter of 1950–51

I was so glad when Guadalcanal came along. Because then the old-timers would have to stop talking about "how it was in Nicaragua." Then we got so sick of hearing about the "Canal." Then it was Iwo and Okinawa. I was glad last year when the Yalu came along and now I'm fed up with the Yalu and how cold it was and the chinks on their Mongol ponies blowing bugles in the damned night. I wish we'd have another decent scrap so we could put the Yalu back where it belongs. With Iwo and the "Canal" and Nicaragua. In the history books.

MARINE COL. GREGORY (1952)
In conversation with a group of his new officers, in Korea
quoted in JAMES BRADY, *THE COLDEST WAR:
A MEMOIR OF KOREA* (1990)

Uxbridge: "By God, sir, I've lost my leg!"
Wellington: "By God, sir, so you have!"

> THE EARL OF UXBRIDGE AND THE DUKE OF WELLINGTON
> At the Battle of Waterloo, June 1815

Never volunteer for anything.

> ANONYMOUS
> OLD ARMY SAYING

Stay with me, God. The night is dark,
The night is cold: my little spark
Of courage dies. The night is long;
Be with me, God, and make me strong.

> "A SOLDIER—HIS PRAYER," *POEMS FROM THE DESERT* (1944)
> British soldier's poem found on a scrap of paper in a
> trench in Tunisia during the Battle of El Agheila by
> members of the British Eighth Army during the African
> campaigns in WWII

Gung Ho!

MARINE COL. EVANS FORDYCE CARLSON
Motto of "Carlson's Raiders," 2nd Raider Battalion (1942)

———————

At the core, the American citizen soldiers knew the difference between right and wrong, and they didn't want to live in a world in which wrong prevailed. So they fought, and won, and we all of us, living and yet to be born, must be forever profoundly grateful.

STEPHEN E. AMBROSE
CITIZEN SOLDIERS (1997)
Discussing American GIs in WWII

———————

A man who is good enough to give his blood for his country is good enough to be given a square deal afterwards.

THEODORE ROOSEVELT, SPEECH (1903)

7

'Nam

Words from Beyond the Wall

They faced the most horrific crossfire ever endured by U.S. fighting forces in the field. Ahead of them was the enemy—motivated, accustomed to fighting in the jungle, supported by guerrillas, supplied and reinforced from sanctuaries that could not be attacked. Behind them was an apathetic nation—at *best* apathetic, sometimes even seemingly hostile by giving ear and succor to those who supported the causes being espoused by the enemy.

Today, looking back, I don't think there are many who doubt that the men and women who served in Vietnam received the shabbiest treatment ever meted out by the American people and their elected representatives. No other troops in our nation's history have faced such slander and back-stabbing. And yet, they were there—fighting for their lives—not because they chose the mission, but because they were ordered to go. And ultimately, as every schoolchild knows, those orders can be traced back to decisions by elected officals and their military advisors.

The fact that American servicemen and women fought bravely in Vietnam is given testimony by their blood that has leached into the soil of that tormented country. Testimony of that blood is readily at hand, for those willing to listen, and it is written on two black granite walls that meet at an angle between the Washington Monument and the Lincoln

Memorial on the peaceful banks of the Potomac River in our nation's capital.

The names of the 58,209 Americans inscribed on the Vietnam Veterans Memorial are powerful reminders of the extreme sacrifices a relative few Americans were making while their fellow citizens lived on in peace and prosperity, attending classes, getting married and raising and educating their children, pursuing careers and personal interests.

Every year now, 2.5 million people come to the Memorial. They touch the names, trace them sometimes on bits of paper. Some of them kneel, some pray, most cry. Some leave little gifts: a picture, a note, flowers, teddy bears.

The Wall was built in 1982, and the statue of the three servicemen was added in 1984, the same year that the Vietnam Veterans Memorial was placed under the stewardship of the National Park Service.

The statue of the Vietnam Women's Memorial was added in 1993.

On its internet site, the Vietnam Veterans Memorial Fund provides interesting commentary on all aspects of the Memorial. (The site is: www.vvmf.org/wall.htm.) Of the statue of the three servicemen, they report: "Composed of three men carrying infantry weapons, the statue grouping has been called both The Three Fighting Men and The Three Servicemen. The men are wearing Vietnam era uniforms and could be from any branch of the U.S. military at that time. Interpretations of the work [by Frederick Hart] vary widely. Some say the troops have the 'thousand yard stare' of combat soldiers. Others say the troops are on patrol and begin looking for their own names as they come upon the Memorial."

The statue of the Vietnam Women's Memorial was created by sculptor Glenna Goodacre, after the inspiration and tireless work of former Army nurse Diane Carlson Evans, whose vision it was to achieve special recognition for the service of women in Vietnam, especially the nurses in combat zones who risked their lives, and sometimes lost their lives, caring for the wounded and dying. [For the younger generations: Unlike "Desert Storm" in 1991 and the military today, women did not serve in the combat arms during the Vietnam War. Most casualties involving women occurred among nurses in forward areas, although service anywhere in the country was dangerous.]

The Vietnam War was the longest in America's history. The first casualties were two American advisors killed on July 8, 1959; the last on May 15, 1975, in connection with the Mayaquez incident. About 2.7 million Americans served in the war zone. Casualties included 300,000 wounded.

Remembrance of Vietnam is not on the wane; it is on the ascendancy. The number of visitors to the Memorial keeps growing. There are many excellent Web sites on the Internet, television documentaries, and many outstanding books, some of which we are fortunate to quote here.

Some of the Internet sites publish letters, poems, and essays written in tribute to Vietnam veterans. On one—www.the wall-usa.com/—Racheline Maltese had this to say after a visit to the Memorial:

"I am only 21. I do not remember the war when it was happening. I did not learn about it in school. To see these men and women with their shirts and flags shakes me. Seeing the things people have left here shakes me. A picture of Jimi Hendrix, a bottle of Seagrams 7, a pack of cigarettes have reduced me to tears.

"I wonder if you [the inscribed veterans] watch us. If you'd like to say thank you for these gifts. I wonder if we mourn for you or for ourselves."

Look deeply into my black granite face and see yourself in the reflection—your face, superimposed on names. Never forget the names, the names, the names—for they hold the answer.

TERRENCE DONNELL (USAF)
Reflections on the Vietnam Memorial Wall

All the wrong people remember Vietnam. I think all the people who remember it should forget it, and all the people who forgot it should remember it.

MICHAEL HERR
THE LONDON OBSERVER (1989)

"Cherries," we were called, and "Newbies" and "FNGs"—"Fuckin' New Guys"—by troops hardened and made less sanguine, perhaps, by just a few months in the bush.

> WILLIAM BROYLES, JR.
> Introduction to *DEAR AMERICA: LETTERS HOME FROM VIETNAM*, BERNARD EDELMAN, ED. (1985)

Well, I'll be dipped in shit—new meat! Sorry bout that boys—"sin loi" buddy . . . you gonna love the Nam, man, for-fucking-ever.

> VIETNAM VETERANS, in MOVIE *PLATOON* (1988)
> Boarding their transport home, taunting new Army troops arriving for combat duty, 1967

In Country

> COMMON EXPRESSION
> Used instead of saying "in Vietnam" or "here"

Aw, jungle's okay. If you know her you can live in her real good, if you don't she'll take you down in an hour. Under.

A "GRUNT" IN VIETNAM
quoted in MICHAEL HERR, *DISPATCHES* (1977)

You will kill ten of our men, and we will kill one of yours, and in the end it will be you who tires of it.

ATTRIBUTED TO HO CHI MINH (1969)

Though our homes are safe from mortars and our countryside from snipers, controversy ranges over the war. It's the warmongers against the peaceniks. Caught in the middle are the returning warriors. While I avoid discussing it, inside I scream . . . I'd like them [the arguing "warmongers and "peaceniks"] to get off their self-righteous asses and learn about war firsthand. I want them to be terrified for their lives day in and day out, to watch a couple of buddies get blown to pieces and *then* see how long they can hang on to their high-and-mighty ideals.

WINNIE SMITH, U.S. ARMY NURSE
quoted in *VOICES FROM VIETNAM*, BARRY DENENBERG, ED.
(1995)

There was never any word as to who the enemy was.

LIEUTENANT WILLIAM CALLEY
At his court martial for the My Lai Massacre in Vietnam,
1971

From our vantage point Vietnam looked very big and very green with its thick covering of jungle. It looked like a great place to have a guerrilla war, if you were going to be the guerrilla.

ROBERT MASON
CHICKENHAWK (1983)
Describing his experience as an assault helicopter pilot with the First Cavalry in Vietnam, 1965–66

Sure, Vietnam is a dirty war. I've never heard of a clean one.

BOB HOPE (1967)

My solution to the problem would be to tell them [the North Vietnamese] that they've got to draw in their horns, or we're going to bomb them back into the Stone Age.

GEN. CURTIS LEMAY
MISSION WITH LEMAY (1965)

Curtis LeMay wants to bomb Hanoi and Haiphong. You know how he likes to go around bombing.

PRESIDENT LYNDON B. JOHNSON
During the Vietnam War

I don't know what it will take out there—500 casualties maybe, maybe 500,000. It's the aughts that scare me.

PRESIDENT LYNDON B. JOHNSON
During the Vietnam War

One of my men turned to me and pointed a hand, filled with cuts and scratches, at a rather distinguished-looking plant with soft red flowers . . . It is a country of thorns and cuts, of guns and marauding, of little hope and great failure. Yet in the midst of it all, a beautiful thought, gesture, and even person can arise among it waving bravely at the death that pours down upon it . . . The flower will also live in the memory of a tired, wet Marine, and has thus achieved a sort of immortality.

LT. MARION LEE KEMPNER (NOV. 1966)
quoted in *DEAR AMERICA: LETTERS HOME FROM VIETNAM* (1985)
This letter to his great-aunt was written less than three weeks before Kempner was wounded by shrapnel in a mine explosion. He ordered the corpsman to care for another wounded man first and died on a medevac en route to the hospital. He was 24 years old.

If you are going to kill someone, you better have a good reason for it. And if you have a good reason, then you better not play around with the killing. We didn't seem to have the good reason, and we were playing around with the killing . . . If I had to be a killer, I wanted to know why I was killing; and the facts didn't match the rhetoric coming out of Washington.

GENERAL CHUCK HORNER
EVERY MAN A TIGER (1999)
Air Force pilot in Vietnam and later Air Force
Commander during the Gulf War

I love the smell of napalm in the morning. It smells like . . . victory.

ROBERT DUVALL, AS LT. COL. KILGORE
Fictional character in the film APOCALYPSE NOW (1979)

Then up this shitting hill we go again. We got to the top and there was a trail watcher there and Brock, the pointman, got twelve rounds right through the gut. That's when Rick, the slack man, got two or three rounds off . . . and his '16 jammed. All I heard was the twelve rounds of the AK and the two from the '16. Rick was up there cussing and banging his M16 against a tree.

BRYAN GOOD
quoted in JAMES R. EBERT, *A LIFE IN A YEAR: THE AMERICAN INFANTRYMAN IN VIETNAM, 1965–1972* (1993)
Describing life as an infantryman during the Vietnam War

"I don't know if you know this," Egan whispered, "but this is one good fuckin company. Dinks seldom hit good companies. They like to pick on the noisy ones where the guys aren't tryin ta stay alive."

FICTIONAL "GRUNT"
JOHN M. DEL VECCHIO, *THE 13TH VALLEY* (1982)

Everything rotted and corroded quickly over there: bodies, boot leather, canvas, metal, morals. Scorched by the sun, wracked by the wind and rain of the monsoon, fighting in alien swamps and jungles, our humanity rubbed off of us as the protective bluing rubbed off the barrels of our rifles.

PHILIP CAPUTO
A RUMOR OF WAR (1977)
On his experiences as a Marine infantry officer in Vietnam, 1965–66

As I was going over there I felt that there was at least a reason for being there. When I got there, I found out there really wasn't; and as it went on, it got worse . . . The enemy was willing to die, really willing to die for what they believe in—a lot of them anyway—and we weren't . . . All we wanted to do was go home. In my war that's how it was.

DAN KREHBIEL quoted in JAMES R. EBERT,
A LIFE IN A YEAR: THE AMERICAN INFANTRYMAN IN VIETNAM, 1965–1972 (1993).
Describing life as an infantryman during the Vietnam War

The grunts weren't even making it to the trees. They had leapt out, screaming murderously, but now they dropped all around us, dying and dead. The lead ship's rotors still turned, but the men inside did not answer. I saw sand spurt up in front of me as bullets tore into the ground. My stomach tightened to stop them. Our door gunners were firing over the prone grunts at phantoms in the trees.

ROBERT MASON
CHICKENHAWK (1983)
Assault helicopter pilot describing his experiences with
the First Cavalry in Vietnam, 1965–66

Visiting the wounded was gut-wrenching, especially when the casualties came day after day . . . I'd go up to a kid who's just lost his leg and was lying there in shock and in pain, and the first thing he'd do would be to *apologize*. "Sir, I'm sorry. I fucked up. I knew better than to do that." Then he'd want to know, "Did anybody die because of me?" I'd tell him how proud I was of him and that everything would be okay.

GEN. H. NORMAN SCHWARZKOPF
IT DOESN'T TAKE A HERO (1992)
Recalling experiences in Vietnam

None of us was a hero. We would not return to cheering crowds, parades, and the pealing of great cathedral bells. We had done nothing more than endure. We had survived, and that was our only victory.

PHILIP CAPUTO
A RUMOR OF WAR (1977)
On his experiences as a Marine infantry officer in Vietnam, 1965–66

8

The Weight of Command

Leadership and Responsibility, Battles Won and Lost

In one of the most famous photographs and film segments of WWII, General Dwight D. Eisenhower is talking to face-blackened and combat-geared paratroopers of the 101st Airborne Division in the final hours before D-Day, June 6, 1944. Though midnight is rapidly approaching, there is still daylight lingering in the long summer evening, and Ike is clearly enjoying his visit with "the men" just before they load up for their parachute drop into Normandy.

Ike is not talking strategy now, or of the exhaustive planning that has preceded this moment—or the difficult decision he has made to press on with the attack despite the dicey weather. Ike knows that for the time being, the work of the generals and staff officers is finished. Now the troops will take over.

The old military saying that "generals die in bed" may very well be true, but before they hit the sack, sick or well, the generals and other officers have a lot to think about. Theirs are the decisions that will determine whether or not their troops have a literal fighting chance to gain a victory, or fight in some hopeless position of disadvantage before the enemy guns.

Some generals, like Montgomery, favored a slow and methodical approach to conserve lives while gaining a victory. Others, Patton is a prime example, claimed that fast, savage attacks in force saved more lives in the long run, while gaining more victories.

Some generals and officers, however brilliant they may be and popular with the troops, seem incapable of giving the orders that will send men to their deaths. Union General George McClellan is probably a good example. His record shows him reluctant to commit his forces, unwilling to "face the arithmetic" of casualty lists, as Abraham Lincoln said.

"To be a good soldier you must love the army. But to be a good officer you must be willing to order the death of the thing you love. That is . . . a very hard thing to do. No other profession requires it. That is one reason why there are so very few good officers. Although there are many good men."

The speaker is a West-Pointer, General Robert E. Lee, speaking to another West-Pointer, his deputy Lt. Gen. James Longstreet, in the late Michael Shaara's classic novel, *The Killer Angels*, the story of the Battle of Gettysburg. Now leading the Confederate forces, both officers are painfully aware that they have broken their earlier vows to the nation that provided their military education. But both are committed to the Army of Northern Virginia now and are frustrated that for the second day they have failed to break the Union Army's position on Cemetery Ridge. Longstreet has literally begged Lee to swing the Army to the right, cut Meade off from Washington, then fight a defensive battle like Fredericksburg where the Union Army suffered grievous losses in desperate frontal attacks against the entrenched Confederates.

Instead of seeing things Longstreet's way, Lee is contemplating a desperate attack of his own for the third day of the battle, on Friday, July 3, straight up Cemetery Ridge into the teeth of Meade's guns.

Written after years of exhaustive research for authenticity, Michael Shaara's expression of Lee's thoughts on the eve of the climactic day of battle is nothing short of a masterpiece, in my eyes anyway.

"It did not take him very long. He was by nature a decisive man, and although this was one of the great decisions of his life, and he knew it, he made it quickly and did not agonize over it. He did not think of the men who would die; he had learned long ago not to do that. The men came here ready to die for what they believed in, for their homes and their honor, and although it was often a terrible death it was always an honorable death.

"He could not retreat now. It might be the clever thing to do, but cleverness did not win victories; the bright combinations rarely worked. You won because the men thought they would win, attacked with courage, attacked with faith, and it was the faith more than anything else you had to protect; that was one thing that was in your hands, and so you could not ask them to leave the field to the enemy."

The following day would not be kind to Robert E. Lee and his men. The attack would fail; he would take full responsibility.

Make no mistake: This is a tough, dirty business for commanders. And you'd better be right. Then you can die in bed.

———

I tell you . . . as officers, that you neither eat, nor drink, nor smoke, nor sit down, nor lean against a tree, until you have personally seen that your men have first had a chance to do those things. If you do this for them, they will follow you to the ends of the earth. And if you do not, I will bust you in front of your regiments.

FIELD MARSHAL WILLIAM JOSEPH SLIM
To a group of his British officers during the campaigns in Burma during WWII

The general who advances without coveting fame and retreats without fearing the disgrace, whose only thought is to protect his country and do good service to his sovereign, is the jewel of the kingdom.

SUN TZU
THE ART OF WAR (C. 490 B.C.)

Great blunders are often made, like large ropes, of a multitude of fibres.

COSETTE, FICTIONAL CHARACTER
VICTOR HUGO
LES MISÉRABLES (1862)

No one can guarantee success in war, but only deserve it.

WINSTON CHURCHILL
THEIR FINEST HOUR (1949)

The military mind always imagines that the next war will be on the same lines as the last. That has never been the case and never will be.

MARSHAL OF FRANCE FERDINAND FOCH (1851–1929)

Comrades, you have lost a good captain to make a bad general.

SATURNINUS (100 B.C.)

Therefore I say: "Know the enemy and know yourself; in a hundred battles you will never be in peril."

SUN TZU
THE ART OF WAR (C. 490 B.C.)

The war will be won or lost on the beaches. We'll have only one chance to stop the enemy and that's while he's in the water . . . struggling to get ashore. Reserves will never get up to the point of attack and it's foolish even to consider them . . . everything we have must be on the coast.

> FIELD MARSHAL ERWIN ROMMEL
> quoted in CORNELIUS RYAN, *THE LONGEST DAY* (1959)
> To his aide, while inspecting the construction of
> shore defenses for Hitler's vaunted
> "Atlantic Wall," April 22, 1944"

———

The whole art of war consists in getting at what is on the other side of the hill, or, in other words, in learning what we do not know from what we do know.

> THE DUKE OF WELLINGTON (ARTHUR WELLESLEY), 1845

Fires are raging aboard the *Kaga*, *Soryu*, and *Akagi* resulting from attacks carried out by land-based and carrier-based attack planes. We plan to have the *Hiryu* engage the enemy carriers. In the meantime, we are returning to the north, and assembling our forces.

> REAR ADMIRAL HIROAKI ABE, SECOND IN COMMAND OF THE JAPANESE MIDWAY STRIKING FORCE
> Radio message to Admiral Isoroku Yamamoto, in effect, announcing the greatest military defeat in Japanese history, June 4, 1942

Brad, this time the Kraut's stuck his head in a meat-grinder. And this time I've got hold of the handle.

> GEN. GEORGE S. PATTON, JR.
> quoted in OMAR N. BRADLEY, *A SOLDIER'S STORY* (1951)
> To Gen. Bradley when told to make the relief of Bastogne a top priority for his Third Army during the Battle of the Bulge, Nov. 1944

I don't mind being called tough, because in this racket it's the tough guys who lead the survivors.

> GENERAL CURTIS LEMAY, STRATEGIC AIR COMMAND (1950s)

The great point to remember is that we are going to finish with this chap Rommel once and for all. It will be quite easy. There is no doubt about it. He is definitely a nuisance. Therefore, we will hit him a crack and finish with him.

> BRITISH FIELD MARSHAL BERNARD LAW MONTGOMERY
> In Africa (early 1940s)

He has lost his left arm, but I have lost my right arm.

> GENERAL ROBERT E. LEE
> Upon learning that Lt. Gen. Thomas J. Jackson
> ("Stonewall Jackson") had suffered an amputation after
> being mistakenly shot by his own pickets at
> Chancellorsville, Virginia, 1863

When soldiers run away in war they never blame themselves: they blame their general or their fellow-soldiers.

DEMOSTHENES (384–322 B.C.)

It is an approved maxim in war, never to do what the enemy wishes you to do, for this reason alone, that he desires it. A field of battle, therefore, which he has previously studied and reconnoitered, should be avoided, and double care should be taken where he has had time to fortify or entrench. One consequence deducible from this principle is, never to attack a position in front, which you can gain by turning.

NAPOLEON I (1769–1821)
MAXIM XVI

Our Army held the war in the hollow of their hand and they would not close it.

ABRAHAM LINCOLN
On General George Meade's failure to pursue Lee's retreating army after Gettysburg, July 1863

I was deeply mortified by the escape of Lee across the Potomac . . . because I believed that General Meade and his noble army had expended all the skill, and toil, and blood, up to the ripe harvest, and then let the crop go to waste.

ABRAHAM LINCOLN
Letter, July 21, 1863

I am quite positive we must give the order . . . I don't like it, but there it is . . . I don't see how we can do anything else.

> GEN. DWIGHT D. EISENHOWER
> Decision to "Go" that committed Allied Forces for D-Day to take place on Tuesday, June 6, 1944 despite the bad weather. The attack had already been postponed from Monday, June 5.

Our landings in the Cherbourg-Havre area have failed to gain a satisfactory foothold and I have withdrawn the troops. My decision to attack at this time and place was based upon the best information available. The troops, the air and Navy did all that bravery and devotion to duty could do. If there is any blame or fault attached to the attempt, it is mine alone.

> GEN. DWIGHT D. EISENHOWER
> quoted in HARRY C. BUTCHER,
> *MY THREE YEARS WITH EISENHOWER*, (1946)
> Message he carried in case of the failure of D-Day, June 6, 1944

Under the command of General Eisenhower, Allied naval forces, supported by strong Allied air forces, began landing Allied armies this morning on the northern coast of France.

> COL. ERNEST DUPUY
> Message the free world had been anticipating for years. The time in London was 0932 in the morning, in Washington 0332 Eastern War Time, June 6, 1944.

The news couldn't be better. As long as they were in Britain we couldn't get at them. Now we have them where we can destroy them.

> ADOLF HITLER
> Responding to news that the Allied invasion had begun, June 6, 1944

After a battle is over people talk a lot about how decisions were methodically reached, but actually there's always a hell of a lot of groping around.

> REAR ADMIRAL JACK FLETCHER
> quoted in WALTER LORD, *INCREDIBLE VICTORY* (1967)
> Discussing the Battle of the Coral Sea, 1942

All warfare is based on deception.

> SUN TZU
> *THE ART OF WAR* (C. 490 B.C.)

There are a lot of people who say that bombing can never win a war. Well, my answer to that is that it has never been tried yet. And we shall see.

> AIR CHIEF MARSHAL ARTHUR HARRIS
> Speaking as Commander in Chief, R.A.F. Bomber Command (1943) quoted in *WORLD AT WAR, NO.* 12, "WHIRLWIND," Syndicated TV Series, Thames

Humility must always be the portion of any man who receives acclaim earned in the blood of his followers and the sacrifices of his friends.

GEN. DWIGHT D. EISENHOWER
Speech, 1945

I think with the Romans, that the general of today should be a solider tomorrow if necessary.

THOMAS JEFFERSON
Letter, 1797

The Creator has not thought proper to mark those in the forehead who are of stuff to make good generals. We are first, therefore, to seek them blindfold, and then let them learn the trade at the expense of great losses.

THOMAS JEFFERSON
Letter, 1813

The good company has no place for the officer who would rather be right than be loved, for the time will quickly come when he walks alone, and in battle no man may succeed in solitude.

BRIG. GEN. S. L. A. MARSHALL (RET.)
MEN AGAINST FIRE (1964)

I shall return.

GENERAL DOUGLAS MACARTHUR
Message on leaving Corregidor in the Philippines for Australia, as ordered by President Roosevelt, March 11, 1942

They died hard—those savage men—not gently like a stricken dove folding its wings in peaceful passing, but like a wounded wolf at bay, with lips curled back in sneering menace, and always a nerveless hand reaching for that long sharp machete knife which long ago they had substituted for the bayonet. And around their necks, as we buried them, would be a thread of dirty string with its dangling crucifix. They were filthy; and they were lousy, and they stank. And I loved them.

GEN. DOUGLAS C. MACARTHUR
Recalling Corregidor
REMINISCENCES

I feel sick at my stomach. I am really low down . . .
They bring in the wounded every minute . . . General Wainwright is a right guy and we are willing to
go on for him, but shells were dropping all night
faster than hell. Damage terrible. Too much for guys
to take. Enemy heavy cross-shelling and bombing.
They have got us all around and from the skies.

CPL. IRVING STROBING
Radio message from Corregidor, May 6, 1942

With broken heart and head bowed in sadness but
not in shame I report to your excellency that today I
must arrange terms for the surrender of the fortified
islands of Manila Bay.

GEN. JONATHAN WAINWRIGHT
Final message from Corregidor, to President Roosevelt,
May 6, 1942

I see that the old flagpole still stands. Have your troops hoist the colors to its peak, and let no enemy ever haul them down.

GEN. DOUGLAS MACARTHUR
Return to Corregidor, March 2, 1945

The military virtue of an Army is, therefore, one of the most important moral powers in War, and where it is wanting, we either see its place supplied by one of the others, such as the great superiority of generalship or popular enthusiasm, or we find the results not commensurate with the exertions made.

GEN. CARL VON CLAUSEWITZ
ON WAR (1832)

Request Lee proceed top speed to cover Leyte; request immediate strike by fast carriers.

ADMIRAL THOMAS C. KINKAID
Urgent dispatch to Admiral William F. Halsey, Jr. ("Bull" Halsey), during the Battle of Leyte Gulf. Halsey had left the San Bernardino Straits unguarded, Oct. 1944.

Help needed from heavy ships immediately . . . Situation critical, battleships and fast carrier strike wanted to prevent enemy penetrating Leyte Gulf.

ADMIRAL THOMAS C. KINKAID
Even after the eventual successful outcome of the battle, arguments would continue to rage over Halsey's move to pursue what was thought to be a "decoy" Japanese fleet.

I have the honour to refer to the very serious calls which have recently been made upon the Home Defence Fighter Units in an attempt to stem the German invasion of the Continent. I hope and believe that our Armies may yet be victorious in France and Belgium, but we have to face the possibility that they may be defeated . . . I must point out that within the last few days the equivalent of ten squadrons have been sent to France, that the Hurricane squadrons remaining in this country are seriously depleted . . . if the Home Defence Force is drained away in desperate attempts to remedy the situation in France, defeat in France will involve the final, complete and irremediable defeat of this country.

AIR CHIEF MARSHAL H. C. T. DOWDING
Appealing to Churchill, who had given the French his word that support for their fight against the Germans would continue, 1940

What General Weygand called the Battle of France is over. I expect the Battle of Britain is about to begin . . . Let us therefore brace ourselves to our duties, and so bear ourselves that, if the British Empire and its Commonwealth last for a thousand years, men will still say, "This was their finest hour."

> WINSTON CHURCHILL
> Speech in the House of Commons upon the fall of France, June 18, 1940

In the early stages of the fight, Mr. Winston Churchill spoke with affectionate raillery of me and my "Chicks" [Britain's fighter aircraft]. He could have said nothing to make me more proud; every Chick was needed before the end.

> AIR CHIEF MARSHAL H. C. T. DOWDING
> Reflecting on the Battle of Britain, 1940

Never in the field of human conflict was so much owed by so many to so few.

> WINSTON CHURCHILL
> Battle of Britain tribute to the Royal Air Force, House of Commons, Aug. 20, 1940

Adversity reveals the genius of a general; good fortune conceals it.

> HORACE
> *SATIRES* (25 B.C.)

I claim we got a hell of a beating. We got run out of Burma and it is humiliating as hell. I think we ought to find out what caused it, go back and retake it.

> GEN. JOSEPH W. ("VINEGAR JOE") STILLWELL
> In a press conference in India following the retreat from Burma, May 1942

There is a tide in the affairs of men
Which, taken at the flood, leads on to fortune;
Omitted, all the voyage of their life
Is bound in shallows and in miseries.

> BRUTUS
> in WILLIAM SHAKESPEARE
> *JULIUS CAESAR* (1599)

We must take the current when it serves,
Or lose our ventures.

> BRUTUS
> in WILLIAM SHAKESPEARE
> *JULIUS CAESAR* (1599)

. . . it is dangerous to detach large forces for any length of time merely for a trick, because there is always the risk of its being done in vain, and then these forces are wanted at the decisive point. The bitter earnestness of necessity presses so fully into direct action that there is no room for that game. In a word, the pieces on the strategical chess-board want that mobility which is the element of stratagem and subtility.

The conclusion which we draw, is that a correct and penetrating eye is a more necessary and more useful quality for a General than craftiness, although that also does no harm if it does not exist at the expense of necessary qualities of the heart, which is often the case.

GENERAL CARL VON CLAUSEWITZ
ON WAR (1832)

Hard pounding, this, gentlemen; try who can pound the longest.

> THE DUKE OF WELLINGTON
> Battle of Waterloo, June 1815

Stand fast . . . we must not be beat—what will they say in England!

> THE DUKE OF WELLINGTON
> quoted in ELIZABETH LONGFORD, WELLINGTON: THE YEARS OF THE SWORD (1969)
> Battle of Waterloo, June 1815

It was about this moment that a distant line of bayonets gleamed on the heights in the direction of Frischemont.

That was the culminating point in this stupendous drama.

The awful mistake of Napoleon is well known. Grouchy expected, Blucher arriving. Death instead of life.

Fate has these turns; the throne of the world was expected; it was Saint Helena that was seen.

VICTOR HUGO
LES MISÉRABLES (1862)
Describing the Battle of Waterloo

It has been a damned serious business . . . the nearest run thing you ever saw in your life.

THE DUKE OF WELLINGTON
quoted in *The Creevey Papers*, (1903)
On the Battle of Waterloo, June 1815

Was it possible for Napoleon to win that battle [Waterloo]. We answer No. Why? Because of Wellington? Because of Blucher? No. Because of God.

VICTOR HUGO
LES MISÉRABLES (1862)

It was time that this vast man [Napoleon] should fall. The excessive weight of this man in human destiny disturbed the balance. The individual alone counted for more than the universal group . . . The moment had arrived for the incorruptible and supreme equity to alter its plan . . . Smoking blood, overcrowded cemeteries, mothers in tears—these are powerful pleaders. When the earth is suffering from too heavy a burden, there are mysterious groanings of the shades, to which the abyss lends an ear.

VICTOR HUGO
LES MISÉRABLES (1862)

Napoleon had been denounced in the infinite, and his fall had been decided on. He embarrassed God.

> VICTOR HUGO
> *LES MISÉRABLES* (1862)

The Battle of Waterloo was won on the playing fields of Eton.

> THE DUKE OF WELLINGTON
> Remark often attributed to the Duke of Wellington but thought by many historians to be apocryphal, even though Wellington had attended Eton as a boy.

The Battle of Yorktown was lost on the playing fields of Eton.

> H. ALLEN SMITH, AMERICAN HUMORIST
> In a quip citing victory in the American Revolution and aimed at the British fondness for the Wellington quote.

The first qualification in a general-in-chief is a cool head—that is, a head which receives just impressions, and estimates things and objects at their real value. He must not allow himself to be elated by good news, or depressed by bad.

The impressions he receives, either successively or simultaneously in the course of the day, should be so classed as to take up only the exact place in his mind which they deserve to occupy; since it is upon a just comparison and consideration of the weight due to different impressions that power of reasoning and of right judgment depends.

Some men are so physically and morally constituted as to see everything through a highly coloured medium. They raise up a picture in the mind on every slight occasion, and give to every trival occurrence a dramatic interest. But whatever knowledge, or talent, or courage, or other good qualities such men possess, nature has not formed them for the command of armies, or the direction of great military operations.

NAPOLEON I (1769–1821)
MAXIM LXXIII

The final test of a leader is that he leaves behind him in other men the conviction and the will to carry on . . . The genius of a good leader is to leave behind him a situation which common sense, without the grace of genius, can deal with successfully.

WALTER LIPPMANN
ROOSEVELT HAS GONE (1945)

As Caesar rode through the Velabrum on the day of his Gallic triumph, the axle of his triumphal chariot broke, and he nearly took a toss; but afterwards ascended to the Capitol between two lines of elephants, forty in all, which acted as torch-bearers. In the Pontic triumph one of the decorated wagons, instead of a stage-set representing scenes from the war, like the rest, carried a simple three-word inscription:

I CAME, I SAW, I CONQUERED!

This referred not to the events of the war, like the other inscriptions, but to the speed with which it had been won.

SUETONIUS
THE TWELVE CAESARS
ROBERT GRAVES, TRANS.

I have more confidence in General McClellan [Union Commander George McClellan] than in any man living. I would forsake everything and follow him to the ends of the earth. I would lay down my life for him.

CAPT. GEORGE ARMSTONG CUSTER
Letter to his parents, 1862

If General McClellan does not want to use the army, I would like to borrow it, provided I could see how it could be made to do something.

ABRAHAM LINCOLN
Comment, 1862

McClellan's vice . . . was always waiting to have everything just as he wanted before he would attack, and before he could get things arranged as he wanted them, the enemy pounced on him.

UNION GEN. GEORGE G. MEADE (C. 1863)

The country will not fail to note—is noting—that the present hesitation to move upon an intrenched enemy, is but the story of Manassas repeated. I beg to assure you that I have never written you, or spoken to you, in greater kindness of feeling than now, nor with a fuller purpose to sustain you. *But you must act.*

ABRAHAM LINCOLN
Letter to McClellan, 1862

Are you not over-cautious when you assume that you can not do what the enemy is constantly doing?

> ABRAHAM LINCOLN
> Letter to McClellan, 1862

I have just read your dispatch about sore tongued and fatigued horses. Will you pardon me for asking what the horses of yours have done since the battle of Antietam that fatigue anything?

> ABRAHAM LINCOLN
> Telegram to McClellan, Oct. 1862

Alas, for my poor country! I know in my inmost heart she never had a truer servant.

> GEN. GEORGE MCCLELLAN
> On being removed from command, Nov. 1862

Nothing except a battle lost can be half so melancholy as a battle won.

THE DUKE OF WELLINGTON
Dispatch, 1815

Operated on this morning. Diagnosis not yet complete but results seem satisfactory and already exceed expectations.

GENERAL LESLIE R. GROVES
Message to President Truman at the Potsdam Conference
that the first atomic bomb had been tested at
Alamogordo, New Mexico, the day before, July 16, 1945

Results clear cut, successful in all respects. Visible effects greater than Trinity [code word for test bomb at Alamogordo].

MESSAGE FROM THE *ENOLA GAY*, AUG. 6, 1945

As the bomb fell over Hiroshima and exploded, we saw an entire city disappear. I wrote in my log the words, "My God, what have we done?"

CAPT. ROBERT LEWIS, COPILOT OF THE *ENOLA GAY*
Interview for the 10th anniversary of the bombing, NBC-TV, May 19, 1955

———

Arms is a profession that, if its principles are adhered to for success, requires an officer do what he fears may be wrong, and yet, according to military experience, must be done, if success is to be attained.

LT. GEN. THOMAS J. ("STONEWALL") JACKSON
Letter to his wife, 1862

It is well that war is so terrible—we should grow too fond of it.

GENERAL ROBERT E. LEE
Speaking to officers at the Battle of Fredericksburg, 1862

Brad, we've gotten a bridge.

GEN. COURTNEY HODGES, COMMANDER OF FIRST ARMY
Phone conversation with General Omar N. Bradley,
describing the capture of the damaged but intact
Remagen bridge across the Rhine, March 7, 1945

The power which the strong have over the weak, the magistrate over the citizen, the employer over the employed, the educated over the unlettered, the experienced over the confiding, even the clever over the silly—the forbearing or inoffensive use of all this power or authority, or total abstinence from it when the case admits it, will show the gentleman in a plain light. The gentleman does not needlessly and unnecessarily remind an offender of a wrong he may have committed against him. He cannot only forgive, he can forget; and he strives for the nobleness of self and mildness of character which impart sufficient strength to let the past be but the past. A true gentleman of honor feels humbled himself when he cannot help humbling others.

CONFEDERATE GEN. ROBERT E. LEE
J. WILLIAM JONES, *PERSONAL REMINISCENCES, ANECDOTES, AND LETTERS OF GENERAL ROBERT E. LEE*
Note found in his satchel after his death (1874)

This [the battle for the Remagen Bridge] was not the biggest battle that ever was, but for me it always typified one thing—the dash, the ingenuity, the readiness of the first opportunity that characterizes the American soldier.

> GEN. DWIGHT D. EISENHOWER
> On the tenth anniversary of the battle of the Remagen Bridge (1955)

Our job now is to worry. Our job now is to work longer and harder than ever, to be disciplined, be hard-nosed. Do not let the Iraqis up off the floor. Kick the shit out of them.

> GEN. CHUCK HORNER
> *EVERY MAN A TIGER* (1999)
> Air Force Commander during the Gulf War, speaking to his staff after hearing of excellent early bombing results on the first night of the Coalition attack to drive Saddam Hussein from Kuwait, Jan. 17, 1991

At midnight I went back to my office. I felt as if I were standing at a craps table in some kind of dream—I'd bet my fortune, thrown the dice, and now watched as they tumbled through the air in slow motion onto the green felt. Nothing I could do would change the way they landed. I sat down and did what soldiers going to war do: I wrote my family saying how much I loved them.

> GEN. H. NORMAN SCHWARZKOPF
> *IT DOESN'T TAKE A HERO* (1992)
> His thoughts after unleashing the Coalition attack to drive
> Saddam Hussein's forces out of Kuwait, Jan. 17, 1991

If you had called him a bastard, that would be one thing. But you called him a *British* bastard. For that, I'm sending you home.

> GEN. DWIGHT D. EISENHOWER
> Addressing an American officer, in England, 1944,
> during the planning for the Normandy Invasion

The frontiers of states or either large rivers, or chains of mountains, or deserts. Of all these obstacles to the march of an army, the most difficult to overcome is the desert; mountains come next, and large rivers occupy the third place.

NAPOLEON I
MAXIMS I

Remember, God provides the best camouflage several hours out of every twenty-four.

GEN. DAVID M. SHOUP, MARINE CORPS COMMANDANT
quoted in *THE NEW YORK TIMES* (1970)

Three-fourths of those things upon which action in War must be calculated, are hidden more or less in the clouds of great uncertainty. Here, then, above all, a fine and penetrating mind is called for, to search out the truth by the tact of its judgment.

An average intellect may, at one time, perhaps hit upon this truth by accident; an extraordinary courage, at another, many compensate for the want of this tact; but in the majority of cases the average result will always bring to light the deficient under-standing.

GEN. KARL VON CLAUSEWITZ
ON WAR (1832)

In forming the plan of a campaign, it is requisite to foresee everything the enemy may do, and to be prepared with the necessary means to counteract it.

Plans of campaign may be modified *ad infinitum* according to circumstances, the genius of the general, the character of the troops, and the features of the country.

NAPOLEON I
MAXIMS II

A general-in-chief should ask himself frequently in the day, What should I do if the enemy's army appeared now in my front, or on my right, or on my left? If he had any difficulty in answering these questions he is ill posted, and should seek to remedy it.

NAPOLEON I
MAXIMS VIII

The transition from the defensive to the offensive is one of the most delicate operations in war.

NAPOLEON I
MAXIMS XIX

———•••••———

It may be laid down as a principle, that the line of operations should not be abandoned; but it is one of the most skillful manoeuvres in war to know how to change it, when circumstances authorise or render this necessary. An army which changes skillfully its line of operation deceives the enemy, who becomes ignorant where to look for its rear, or upon what weak points it is assailable.

NAPOLEON I
MAXIMS XX

When you are occupying a position which the enemy threatens to surround, collect all your force immediately, and menace him with an offensive movement. By this manoeuvre you will prevent him from detaching and annoying your flanks, in case you should judge it necessary to retire.

NAPOLEON I
MAXIMS XXIII

When you have resolved to fight a battle, collect your whole force. Dispense with nothing. A single battalion sometimes decides the day.

NAPOLEON I
MAXIMS XXIX

War on paper and war in the field are as different as darkness from light, fire from water, or heaven from earth.

WILLIAM FALKNER
THE LITTLE BRICK CHURCH (1882)

Infantry, cavalry, and artillery are nothing without each other. They should always be so disposed in cantonments as to assist each other in case of surprise.

NAPOLEON I
MAXIMS XLVII

All the planning and the thousands of actions that go on in war depend on faith and trust. No single commander can know all that needs to be known, can be everywhere to make every decision that needs to be made, or can direct every action that is taken.

GEN. CHUCK HORNER (Ret.)
EVERY MAN A TIGER (1999)

When near, make it appear that you are far away; when far away, that you are near.

SUN TZU
THE ART OF WAR (C. 490 B.C.)

Keep him under a strain and wear him down.

SUN TZU
THE ART OF WAR (C. 490 B.C.)

Thus, while we have heard of blundering swiftness in war, we have not yet seen a clever operation that was prolonged.

> SUN TZU
> *THE ART OF WAR* (C. 490 B.C.)

When the strike of a hawk breaks the body of its prey, it is because of timing.

> SUN TZU
> *THE ART OF WAR* (C. 490 B.C.)

In the tumult and uproar the battle seems chaotic, but there is no disorder; the troops appear to be milling about in circles but cannot be defeated.

> SUN TZU
> *THE ART OF WAR* (C. 490 B.C.)

Order or disorder depends on organization; courage or cowardice on circumstances; strength or weakness on dispositions.

SUN TZU
THE ART OF WAR (C. 490 B.C.)

Sometimes I use light troops and vigorous horsemen to attack where he is unprepared, sometimes strong crossbowmen and bow-stretching archers to snatch key positions, to stir up his left, overrun his right, alarm him to the front, and strike suddenly into his rear.

In broad daylight I deceive him by the use of flags and banners and at night confuse him by beating drums. Then in fear and trembling he will divide his forces to take precautionary measures.

TU MU
quoted in *ART OF WAR*, SUN TZU (490 B.C.)

Show him there is a road to safety, and so create in his mind the idea that there is an alternative to death. Then strike.

> Tu Mu
> quoted in *Art of War*, Sun Tzu (490 B.C.)

Hence that general is skilful in attack whose opponent does not know what to defend; and he is skilful in defense whose opponent does not know what to attack.

> Sun Tzu
> *The Art of War* (C. 490 B.C.)

9

The Fallen

When an editor like myself sits down to say a few words about our dead and wounded, brevity is certainly the order of the day.

I can tell you that I loved them and treasure the memory of what they did, but my words seem pale and limp beside the stirring tributes produced by more talented spokesmen and witnesses who were there at the time of loss.

Hopefully, the quotes that follow will seem as poignant to you as they do to me. And even though we did not say or write the words, we feel them as deeply as if they had been our own.

"Not in vain" may be the pride of those who have survived and the epitaph of those who fell.

WINSTON CHURCHILL
Addressing the House of Commons, 1944

Your death will not prevent future wars, will not make the world safe for your children. Your death means no more than if you had died in your bed, full of years and respectability, having begotten a tribe of young. Yet by your courage in tribulation, by your cheerfulness before the dirty devices of this world, you have won the love of those who have watched you. All we remember is your living face, and that we loved you for being of our clay and our spirit.

GUY CHAPMAN, BRITISH WWI OFFICER
A PASSIONATE PRODIGALITY (1965)

Pile the bodies high at Austerlitz and Waterloo.
Shovel them under and let me work—
   I am grass; I cover all.

And pile them high at Gettysburg
And pile them high at Ypres and Verdun.
Shovel them under and let me work.

CARL SANDBURG
"GRASS" (1918)

The last war [World War One], during the years 1915, 1916, 1917, was the most colossal, murderous, mismanaged butchery that has ever taken place on earth.

ERNEST HEMINGWAY
MEN AT WAR (1942)

"This land here cost twenty lives a foot that summer . . . See that little stream—we could walk to it in two minutes. It took the British a month to walk to it—a whole empire walking very slowly, dying in front and pushing forward behind. And another empire walked very slowly backward a few inches a day, leaving the dead like a million bloody rugs. No European will ever do that again in this generation."

DICK DIVER, FICTIONAL CHARACTER
TENDER IS THE NIGHT, F. SCOTT FITZGERALD (1933)
Describing the Somme battlefield

Tell those at Thermopylae that here in obedience to her laws we lie.

HERODOTUS (484–424 B.C.)
Inscription on the battlefield where 300 Spartan soldiers lost their lives against overwhelming numbers of Persian forces in Greece in 480 B.C. Under Leonidas I, the Spartans made the battle of Thermopylae a legend of courage.

Dear Captain Agius,

I wish to take this opportunity of thanking you for your kind letter of sympathy, and for the few details you were able to give me concerning my dear husband's death . . . It was a great relief to know that dear Harold did not suffer any pain, although what would I not give to have had just one last message from him. We have been married such a short time (only five months) and I cannot realize that he has gone—never to see him again. The last time we were together he was so happy and well and eager to do his level best for his Country at all cost . . .

FLORENCE E. SCARLETT
quoted in LYN MACDONALD, *SOMME* (1983)
Letter to her husband's commanding officer following his death in the Somme campaign (1916)

I have been shown in the files of the War Department a statement of the Adjutant General of Massachusetts, that you are the mother of five sons who have died gloriously on the field of battle.

I feel how weak and fruitless must be any words of mine which should attempt to beguile you from the grief of a loss so overwhelming. But I cannot refrain from tendering to you the consolation that may be found in the thanks of the Republic they died to save.

I pray that our Heavenly Father may assuage the anguish of your bereavement, and leave you only the cherished memory of the loved and lost, and the solemn pride that must be yours, to have laid so costly a sacrifice upon the alter of Freedom.

PRESIDENT ABRAHAM LINCOLN
Letter to Mrs. Lydia Bixley of Massachusetts (1864)

In the midst of life we are in death.
Earth to earth, ashes to ashes, dust to dust; in sure and certain hope of the Resurrection into eternal life.

THE BOOK OF COMMON PRAYER, "THE BURIAL OF THE DEAD" (1662)

What is life? It is the flash of a firefly in the night. It is the breath of a buffalo in the wintertime. It is the little shadow which runs across grass and loses itself in the sunset.

CROWFOOT, BLACKFOOT WARRIOR AND ORATOR (1890)

Sit on the bed. I'm blind and three parts shell.
Be careful; can't shake hands now; never shall.
Both arms have mutinied against me,—brutes.
My fingers fidget like ten idle brats.

I tried to peg out soldierly,—no use!
One dies of war like any old disease.
This bandage feels like pennies on my eyes.
I have my medals?—Discs to make eyes close.
My glorious ribbons?—Ripped from my own back
In scarlet shreds . . .

WILFRED OWEN (1893–1918)

———

What passing-bells for those who die as cattle?
    Only the monstrous anger of the guns.
    Only the stuttering rifles' rapid rattle
Can patter out their hasty orisons.

WILFRED OWEN (1893–1918)

Soldier, rest! Thy warfare o'er,
Dream of fighting fields no more:
Sleep the sleep that knows not breaking,
Morn of toil, nor night of waking.

SIR WALTER SCOTT
THE LADY OF THE LAKE (1810)

"Yes, we're guilty. What are we to do now? What can I do to wash my hands?"

Segal sighed wearily, spoke without exultation or joy or bitterness, speaking not for himself, but for the first Jew brained on a Munich street long ago and the last American brought to earth that afternoon by a sniper's bullet outside Chartres, and for all the years and all the dead and all the agony in between. "You can cut your throat," he said, "and see if the blood will take the stain out."

IRWIN SHAW
SHORT STORIES OF IRWIN SHAW (1966)

How sleep the brave, who sink to rest,
By all their country's wishes blest!

WILLIAM COLLINS
"ODE WRITTEN IN THE YEAR 1776"

If I should die, think only this of me:
    That there's some corner of a foreign field
That is for ever England. There shall be
    In that rich earth a richer dust concealed;
A dust who England bore, shaped, made aware,
    Gave, once, her flowers to love, her ways to roam,
A body of England's, breathing English air,
    Washed by the rivers, blest by suns of home.

RUPERT BROOKE
"THE SOLDIER" (1914)
Brooke lost his life on duty with the Royal Navy in 1915

. . . The world will little note, nor long remember what we say here, but it can never forget what they did here. It is for us the living, rather, to be dedicated here to the unfinished work which they who fought here have thus far so nobly advanced. It is rather for us to be here dedicated to the great task remaining before us—that from these honored dead we take increased devotion to that cause for which they gave the last full measure of devotion—that we here highly resolve that these dead shall not have died in vain—that this nation, under God, shall have a new birth of freedom-and that the government of the people, by the people, for the people, shall not perish from the earth.

PRESIDENT ABRAHAM LINCOLN
THE GETTYSBURG ADDRESS (NOV. 19, 1863)

The Gestapo guards had panicked. Within minutes of the news of the landings [D-Day], two machine guns had been set up in the prison courtyard. In groups of ten the male prisoners were led out, placed against the wall and executed. They had been picked up on a variety of charges, some true, some false . . . ninety-two in all, of whom only forty were members of the French underground. On this day, the day that began the great liberation . . . these men were slaughtered.

CORNELIUS RYAN
THE LONGEST DAY (1959)
On German D-Day atrocities committed
against French civilians, June 6, 1944

The tumult and the shouting dies;
   The captains and the kings depart:
Still stands Thine ancient sacrifice,
   An humble and contrite heart.
Lord God of Hosts, be with us yet,
Lest we forget—lest we forget!

> RUDYARD KIPLING (1865–1936)
> "RECESSIONAL"

Underneath this wooden cross there lies
A Christian killed in battle. You who read,
Remember that this stranger died in pain;
And passing here, if you can lift you eyes
Upon a peace kept by a human creed,
Know that one soldier has not died in vain.

> KARL SHAPIRO
> "ELEGY FOR A DEAD SOLDIER," *V-LETTER AND OTHER POEMS* (1944)

The paths of glory lead but to the grave.

THOMAS GRAY
"ELEGY IN A COUNTRY CHURCHYARD" (1750)

Gentlemen, I rather have written those lines than take Quebec.

BRITISH GEN. JAMES WOLFE
Commenting to his officers on Gray's "Elegy in a
Country Churchyard," just before Wolfe was killed in the
Battle of Quebec (1759)

Then, while it was still summer, the Marine Corps notified me my footlocker had been shipped and I could pick it up at Grand Central Station . . . I handed a paper to an employee of the railroad, and he led us deeper into the cellar. There, along with lost baggage and my footlocker, were the coffins from Korea, stacked and tidy, each with its American flag neatly lashed on.

Like Mack and Simonis and Captain Chafee and me, they too were home.

JAMES BRADY
THE COLDEST WAR, A MEMOIR OF KOREA (1990)
Returning to New York City from Marine duty in
Korea, 1952

In Flanders fields the poppies blow
Between the crosses, row on row,
    That mark our place; and in the sky
    The larks, still bravely singing, fly
Scarce heard amid the guns below

We are the Dead. Short days ago
We lived, felt dawn, saw sunset glow,
    Loved and were loved, and now we lie
        In Flanders fields.

Take up our quarrel with the foe;
To you from failing hands we throw
    The torch; be yours to hold it high.
    If ye break faith with us who die
We shall not sleep, though poppies grow
        In Flanders fields.

        JOHN MCRAE (1872–1918)
        "IN FLANDERS FIELDS"

Over the first four days of the Bulge, when his panzers were on the move, Peiper's [Lt. Col. Jochen Peiper] command murdered approximately 350 American POWs and at least 100 unarmed Belgian civilians. But Peiper was not present at Malmedy while the GIs from Battery B [285th Field Artillery Observation Battalion, 7th Armored Division] were rounded up, about 250 of them. . . . The Germans parked a tank at either end of the snow-covered field near the wood. A command car drew up. The German officer stood up, pulled his pistol, took deliberate aim at an American medical officer in the front frank of the POWs, and shot him. As the doctor fell, the German shot an officer next to him. Then the two tanks opened up with their machine guns on the POWs. . . . Men took off for the wood, and many made it. Of the 150 POWs, 70 survived.

STEPHEN E. AMBROSE
CITIZEN SOLDIERS (1997)
On the Malmedy massacre during the
Battle of the Bulge, Dec. 17, 1944

My sword I give to him that shall succeed me in my prilgrimage, and my courage and skill to him that can get it. My marks and scars I carry with me, to be witness for men, that I have fought his battles who now will be my rewarder.

So he passed over, and all the trumpets sounded for him on the other side.

JOHN BUNYAN
PILGRIM'S PROGRESS (1678)

The only way we got out of Frozen Chosin is because a lot of young guys know how to fight. God bless the Chosin Marines. They are my brothers for life. . . . Every Memorial Day my thoughts go drifting back to those youngsters who never came home. I can still see them as they were then. They'll never grow old.

LIEUTENANT HENRY LITVIN, U.S. NAVY
A Battalion Surgeon with the Marines, Litvin is describing fighting out of the Chosin Reservoir in North Korea, Nov.-Dec., 1950
quoted in MARTIN RUSS, *BREAKOUT* (1999)

10

Final Battles

Sooner or later, all soldiers face a final battle, whether they "die with their boots on" or live to a tranquil or garrulous and crotchety old age.

Sometimes the end comes long after the uniforms have been put away, the old battles and campaigns misty in memories now rich with new reflections of family and friends, careers and personal interests. For other veterans, the battles never fade, and their thoughts turn to them often, even in the delirium of ebbing life.

Truly then, as General MacArthur reminded us, some old soldiers "just fade away." Some, however, wave goodbye as best they can.

For them all, in war and peace, the bugler blows "Taps" in the daytime.

———•◦•◦•———

I am convinced that the best service a retired general can perform is to turn in his tongue along with his suit, and to mothball his opinions.

GEN. OMAR N. BRADLEY
Armed Forces Day Address quoted in
THE *NEW YORK TIMES*, MAY 17, 1959

The broken soldier, kindly bade to stay,
Sat by his fire, and talk'd the night away;
Wept o'er his wounds, or, tales of sorrow done,
Shoulder'd his crutch, and show'd how fields were won.

OLIVER GOLDSMITH
"THE DESERTED VILLAGE" (1770)

Old, unhappy, far-off things,
And battles long ago.

WILLIAM WORDSWORTH
"THE SOLITARY REAPER" (1807)

I love war and responsibility and excitement. Peace
is going to be hell on me.

GEN. GEORGE S. PATTON, JR.
quoted in LADISLAS FARAGO, *PATTON,*
ORDEAL AND TRIUMPH (1964)

Hear me, my chiefs! I am tired; my heart is sick and sad. From where the sun now stands I will fight no more forever.

> CHIEF JOSEPH OF THE NEZ PERCE (1877)
> in DEE BROWN, *BURY MY HEART AT WOUNDED KNEE* (1970)

Here are the final words of one of history's most beloved generals, spoken in delirium, after heart ailments had incapacitated him in late September. First, he is in battle, calling once again on A.P. Hill:

"Tell Hill he must come up."

And then, quietly:

"Strike the tent."

> GEN. ROBERT E. LEE (OCT. 12, 1870)
> quoted in SHELBY FOOTE, *THE CIVIL WAR: A NARRATIVE, RED RIVER TO APPOMATTOX* (1974)

As he fell asleep he had been thinking of the subject that now always occupied his mind—about life and death, chiefly about death. He felt himself nearer to it.

"Love? What is love?" he thought.

"Love hinders death. Love is life. All, everything that I understand, I understand only because I love. Everything is, everything exists, only because I love. Everything is united by it alone. Love is God, and to die means that I, a particle of love, shall return to the general and eternal source."

PRINCE ANDREW BOLKONSKI, FICTIONAL CHARACTER
in LEO TOLSTOY
WAR AND PEACE (1868–69)

After being accidentally shot by his own pickets while inspecting his lines at dusk at Chancellorsville, Virginia, Lt. Gen. Thomas J. Jackson ("Stonewall Jackson") suffered the amputation of his left arm. Jackson's condition worsened. Pneumonia set in, and on Sunday, May 10, 1863, he knew he was dying:

It is the Lord's day, my wish is fulfilled. I have always desired to die on Sunday.

Later, in delirium, Jackson's final words brought him to a soldier's ultimate peace:

Let us cross over the river and rest under the shade of the trees.

> LT. GEN. THOMAS J. JACKSON
> quoted in SHELBY FOOTE, *THE CIVIL WAR, A NARRATIVE,*
> on his deathbed, May 10, 1863
> *FREDERICKSBURG TO MERIDIAN* (1963)

I am closing my fifty-two years of military service. When I joined the Army, even before the turn of the century, it was the fulfillment of all my boyish hopes and dreams. The world has turned over many times since I took the oath on the Plain at West Point, and the hope and dreams have long since vanished, but I still remember the refrain of one of the most popular barrack ballads of that day, which proclaimed most proudly, that "Old soldiers never die—they just fade away." And like the solider of the ballad, I now close my military career and just fade away—an old solider who tried to do his duty as God gave him the light to see that duty. Goodbye.

GEN. DOUGLAS MACARTHUR
Addressing the U.S. House of Representatives after being dismissed as Supreme Commander of the Pacific by President Harry S. Truman (1951)

People grow old only by deserting their ideals. Years may wrinkle the skin, but to give up interest wrinkles the soul . . . You are as young as your faith, as old as your doubt, as young as your self-confidence, as old as your fear, as young as your hope, as old as your despair. In the central place of every heart there is a recording chamber, so long as it receives messages of beauty, hope, cheer, and courage, so long are you young.

GEN. DOUGLAS MACARTHUR
quoted in WILLIAM MANCHESTER, *AMERICAN CAESAR* (1978)

They shall beat their swords into ploughshares,
and their spears into pruning-hooks; nation
shall not lift up sword against nation; neither
shall they learn war any more.

BIBLE, ISAIAH

Nothing will mix and amalgamate more easily than an old priest and an old soldier. In reality, they are the same kind of man. One has devoted himself to his country upon earth, the other to his country in heaven; there is no other difference.

MARIUS, FICTIONAL CHARACTER
in VICTOR HUGO
LES MISÉRABLES (1862)

Don't go back to visit the old front. If you have pictures in your head of something that happened in the night in the mud at Paschendaele or the first wave working up the slope of Vimy, do not try and go back to verify them. It is no good . . . It is like going into the empty gloom of a threater where the char-women are scrubbing.

ERNEST HEMINGWAY
"A VETERAN VISITS THE OLD FRONT,"
TORONTO DAILY STAR (1922)

Day Is Done,
Gone the Sun,
From the Earth,
From the Hill,
From the Sky.
All Is Well,
Safely Rest,
God Is Nigh

FOOTNOTE: THE STORY OF *TAPS*:

"Taps" is thought to be a revision of an old French bugle signal called "Tattoo" that told the troops to cease their evening's drinking and prepare for another call to extinguish fires an hour later. The word "Taps" is an alteration of the obsolete word "taptoo," derived from the Dutch "taptoe," which meant to shut the tap of a keg. During the Civil War, in July, 1862, Union General Daniel Adams Butterfield hummed a version of the old "Tattoo" to an aide, who wrote it down to music. The Brigade bugler, Oliver W. Norton, experimented with the notes until Butterfield was satisfied with the melody. Norton played the new call at the end of each day thereafter, instead of the regulation call, and its popularity was quick to spread—even to Confederate units. Originally intended and used as a "Lights Out" call, the earliest reference to the mandatory use of *Taps* at military funeral ceremonies is found in the U.S. Army Infantry Drill Regulations for 1891, although it undoubtedly was used unofficially long before that time.

EXCERPTED from research by Master Sergeant Jari A. Villanueva, U.S. Air Force Band, available on the following Web site:

www.west-point.org

Authors and Works Quoted

Stephen E. Ambrose
D-Day June 6, 1944 (1994)
Citizen Soldiers (1997)
Wall St. Journal article on Pearl Harbor (1999)

Herbert Asquith
"The Volunteer," *Poems 1912–33* (1934)

Hanson W. Baldwin
Battles Lost and Won (1966)

James Lee Barrett
Green Berets, screenplay based on novel by Robin Moore
(1968)

Francis Bellamy
Original Pledge of Allegiance to the Flag (1892)

Martin Blumenson
Patton (1985)

Napoleon I
Maxims (c. 1831)

Linda Botts, Editor
Loose Talk (1980)

General Omar N. Bradley
A Soldier's Story (1951)

James Brady
The Coldest War, A Memoir of Korea (1990)

Rupert Brooke
"The Soldier" (1914)

William Broyles, Jr.
*Introduction, Dear America: Letters Home
from Vietnam* (1985)

Dee Brown
Bury My Heart at Wounded Knee (1970)

John Bunyan
Pilgrim's Progress (1678)

George H. W. Bush
Speech (1990)

Harry C. Butcher
My Three Years with Eisenhower (1946)

Lord Byron
Don Juan

Julius Caesar
Gallic Wars

Philip Caputo
A Rumor of War (1977)

Bruce Catton
A Stillness at Appomattox (1953)

Iris Chang
The Rape of Nanking: The Forgotten Holocaust of World War II (1997)

Guy Chapman
A Passionate Prodigality (1965)

Geoffrey Chaucer (1343–1400)
The Knight's Tale, translated from Middle English by John Dryden

Tom Clancy, with General Chuck Horner
Every Man a Tiger (1999)

Winston Churchill
Speeches
A Roving Commission ((1930)
Their Finest Hour (1949)

William Collins
"Ode Written in the Year 1776"

George M. Cohan
"Over There" (1917)

Stephen Crane
The Red Badge of Courage (1895)

Robert MacArthur Crawford
Air Force Song (1939)

Sir Edward Creasy
"The Victory of the Americans over Bourgoyne at
Saratoga, 1777," from *Men at War* (1942)

John M. Del Vecchio
The 13th Valley (1982)

Michel De Montaigne (1533–1592)
Essays, Book II

Barry Denenberg, Editor
Voices from Vietnam (1995)

Frederick Douglass
Men of Color, To Arms (1863)

Ronald J. Drez, Editor
Voices of D-Day (1994)

David Douglas Duncan
This Is War (1951)

Rev. John B. Dykes, Co-Author
The Navy Hymn, Original English Hymns (c. 1860–61)

James R. Ebert
A Life in a Year: The American Infantryman in Vietnam, 1965–1972 (1993)

Bernard Edelman, Editor
Dear America: Letters Home from Vietnam (1985)

Ralph Waldo Emerson
Concord Hymn (1837)
Heroism (1841)

Dwight D. Eisenhower
First Inaugural Address (1953)

Bergen Evans, Editor
Dictionary of Quotations (1969)

William Falkner
The Little Brick Church (1882)

Ladislas Farago
Patton, Ordeal and Triumph (1964)

F. Scott Fitzgerald
Tender Is the Night (1933)

Shelby Foote
The Civil War: A Narrative, Fredericksburg to Meridian (1963)
The Civil War: A Narrative, Red River to Appomattox (1974)

Commander Howell Forgy
 ". . . And Pass the Ammunition" (1944)

Benjamin Franklin
 Remarks (C. 1775–76)

Paul Fussell
 The Great War and Modern Memory (1975)

Oliver Goldsmith
 "The Deserted Village" (1770)

Thomas Gray
 "Elegy in a Country Churchyard" (1750)

Giuseppe Garibaldi
 Remarks (1849)

Major Edmund L. Gruber
 The Caisson Song (1908)

Nigel Hamilton
 Monty, The Making of a General (1981)

Private Marion Hargrove
 See Here, Private Hargrove (1942)

Air Chief Marshal Arthur Harris
 Address (1943)
 Message to Air Crews (1943)

Thomas Heggen
 Mister Roberts (1948)

Joseph Heller
  Catch 22 (1961)

Ernest Hemingway
  "Soldier's Home," *In Our Time* (1925)
  "Now I Lay Me," *Men Without Women* (1927)
  A Farewell to Arms (1929)
  Collier's Magazine article on D-Day (1944)
  Men At War, Editor (1942)

Herodotus (484–424 B.C.)
  Histories

Michael Herr, with Gustav Hasford
  Dispatches (1968)
  The London Observer (1989)

Ian V. Hogg,
  Patton (1982)

Horace
  Satires (25 B.C.)

Maurice Horn, Editor
  The World Encyclopedia of Comics (1976)

General Chuck Horner, with Tom Clancy
  Every Man a Tiger (1999)

A. E. Housman
  More Poems, "No. 36" (1936)

Julia Ward Howe
The Battle Hymn of the Republic (1862)

Victor Hugo
Les Misérables (1862)

Thomas Jefferson
Letters (1797 and 1813)

James Jones
From Here to Eternity (1951)
WW II (1975)

Jack Judge, with Harry Williams
The Tipperary Song (1908)

MacKinlay Kantor, with General Curtis E. LeMay
Mission With LeMay (1965)

John Keegan, with Richard Holmes
Soldiers: A History of Men in Battle (1985)

Lt. Marion Lee Kempner
Dear America, Letters Home From Vietnam (1985)

John F. Kennedy
Speech As Presidential Candidate (1960)

Francis Scott Key
The Star Spangled Banner (1814)

Rudyard Kipling
Gunga Din
"Recessional"

Robert S. La Forte, Co-Editor
Remembering Pearl Harbor (1991)

Lt. Sharon A. Lane
Dear America: Letters Home from Vietnam (1985)

Howard Langer, Editor
World War II: An Encyclopedia of Quotations (1999)

General Curtis E. LeMay, with MacKinlay Kantor
Mission with LeMay (1965)

Abraham Lincoln
Speeches, Letters, Dispatches

Livy (59 B.C.–A.D. 17)
*The History of Rome, Book II,
Horatius at the Bridge*

Henry Wadsworth Longfellow
Journal (1861)

Elizabeth Longford
Wellington: Year of the Sword (1969)

Walter Lord
Incredible Victory (1967)

John Milus
Apocalypse Now, screenplay (1979)

Darryl Lyman, Editor
Civil War Quotations (1995)

General Douglas MacArthur
  Inscription for gymnasium portal, U.S. Military
  Academy (c. 1919–22)
  Reminiscences

Thomas Babington Macaulay
  Lays of Ancient Rome (1842)

Lyn Macdonald
  Somme (1983)

John McRae (1872–1918)
  "In Flanders Fields"

Racheline Maltese
  "Reflections on Vietnam Wall," (1999)

William Manchester
  Goodbye, Darkness (1979)
  The Last Lion, Visions Of Glory, 1874–1932 (1983)
  The Last Lion, Alone, 1932–1940 (1988)
  American Caesar (1978)

Roland E. Marcello, Co-Editor
  Remembering Pearl Harbor (1991)

S.L.A. Marshall
  Battle At Best (1963)
  Men Against Fire (1964)

Robert Mason
  Chickenhawk (1983)

Bill Maudlin
  Up Front (1945)

Midshipman Alfred H. Miles
  The Navy March, lyrics (1906)

Thomas Moore (1780–1852)
  "Come, Send Round the Wine"

Samuel Eliot Morison
  *History of United States Naval Operations in World War
  II, Volumes Four and Five* (1949)

Edward R. Murrow
  Remarks (1954)

George Orwell
  The Art of Donald McGill (1941)

Wilfred Owen (1893–1918)
  "Exposure"
  "A Terre"

Tim Page
  Tim Page's Nam (1983)

Thomas Paine
  The American Crisis, No. 1 (1776)

Francis Parkman
  Montcalm and Wolfe (1906)

General George S. Patton Jr.
War As I Knew It (1941)

Marshal Henri Philippe Petain
Remarks (1916)

Peter Petre, with General H. Norman Schwarzkopf
It Doesn't Take a Hero (1992)

Plutarch
Lives

Ernie Ply
Brave Men (1943)

General Colin Powell
In His Own Words (1995)

Gordon W. Prange
At Dawn We Slept (1981)

Harry W. Pfanz
Gettysburg, The Second Day (1987)

Michael Reynolds
Hemingway, The Final Years (1999)

Mary Roberts Rinehart
Kings, Queens and Pawns (1915)

Derek Robinson
Piece Of Cake (1983)

Robert Rodat
    Saving Private Ryan, screenplay (1998)

Franklin Delano Roosevelt
    Pearl Harbor Speech (1941)

Colonel Theodore Roosevelt
    "The Sword of the Lord and Of Gideon," Rank and File (1928)

Bill D. Ross
    Iwo Jima: Legacy Of Valor (1985)

Martin Russ
    Breakout (1999)

Cornelius Ryan
    The Longest Day (1959)
    A Bridge Too Far (1977)

James Salter
    The Hunters (1956, republished 1997)

Carl Sandburg
    "Grass" (1918)

Arthur Schlesinger
    A Thousand Days (1965)

General H. Norman Schwarzkopf, with Peter Petre
    It Doesn't Take a Hero (1992)

Sir Walter Scott
    Count Robert Of Paris (1862)

The Lady of the Lake (1810)

William Shakespeare
Anthony and Cleopatra
Julius Caesar (1599)
Hamlet (1601)
King Henry the Fifth (1599)
King John (1596)
King Richard the Third (1593)

Karl Shapiro
"Elegy For a Dead Soldier,"V-Letter and Other Poems (1944)

Irwin Shaw
Short Stories of Irwin Shaw (1966)

Robert Gould Shaw
Monument Inscription, Boston Commons (1897)

Lisa Shaw, Editor
General Colin Powell: In His Own Words (1995)

Robert Sherrod
Tarawa (1944)

Hugh Sidey
Time magazine (1985)

Winnie Smith
Voices from Vietnam (1995)

Colonel Charles E. Stanton
    Speech (1917)

Bert Stiles
    Serenade to the Big Bird (1947)

Suetonius
    Lives of the Twelve Caesars, translated by
    Robert Graves

Gay Talese
    Fame and Obscurity: Portraits (1970)

Tacitus (A.D. 54–119)
    The Histories

Alfred Lord Tennyson
    The Charge of the Light Brigade (1854)

Thucydides
    Peloponnesian War, Book II (c. 471–401 B.C.)

John Toland
    Battle: The Story of the Bulge (1959)
    The Rising Sun (1970)

Leo Tolstoy
    War and Peace (1868–69)

Peter Townsend
    Duel Of Eagles (1970)

Peter B. Tsouras, Editor
Military Quotations From the Civil War, In the Words of the Commanders (1998)

Sun Tzu
The Art of War (c. 500 BC)

Frederick F. Van de Water
Glory Hunter (1934)

Master Sergeant Jari A. Villanueva
"Story of *Taps*"

Voltaire
Remarks (1770)

General Karl von Clausewitz
On War (1832)

Harry Williams, with Jack Judge
The Tipperary Song (1908)

William L. White
They Were Expendable (1945)

Rev. William Whiting, Co-Author
The Navy Hymn, Original English Hymns (c. 1860–61)

William Wordsworth
"The Solitary Reaper" (1770)

Herman Wouk
War And Remembrance (1978)

General Chuck Yeager
  Yeager (1985)

Lt. Charles A. Zimmerman
  The Navy March, music portion (1906)

INDEX